Lecture Notes in Mathematics

Edited by A. Dold and B. Eckmann
Series: Mathematisches Institut der Universität Bonn
Adviser: F. Hirzebruch

603

Boris Moishezon

Complex Surfaces and Connected Sums of Complex Projective Planes

---

Springer-Verlag
Berlin Heidelberg New York 1977

Author
Boris Moishezon
Mathematics Department
Tel-Aviv University
Tel-Aviv/Israel

Library of Congress Cataloging in Publication Data

Moishezon, Boris, 1937-
Complex surfaces and connected sums of complex projective planes.

(Lecture notes in mathematics ; 603)
Bibliography: p.
Includes index.
1. Surfaces, Algebraic. 2. Projective planes.
3. Manifolds (Mathematics) I. Title. II. Series:
Lecture notes in mathematics (Berlin) ; 603.
QA3.L28 no. 603 [QA571] 510'.8s [516'.352] 77-22136

AMS Subject Classifications (1970): 14J99, 32J15, 57A15, 57D55

ISBN 3-540-08355-3 Springer-Verlag Berlin Heidelberg New York
ISBN 0-387-08355-3 Springer-Verlag New York Heidelberg Berlin

Printed in Germany
Printing and binding: Beltz Offsetdruck, Hemsbach/Bergstr.
2141/3140-543210

TABLE OF CONTENTS

Introduction

In [1] Wall proved the following theorem:

If V_1, V_2 are simply-connected compact 4-manifolds, which are h-cobordant to each other, then there exists an integer $k \geq 0$ such that $V_1 \# k(S^2 \times S^2)$ is diffeomorphic to $V_2 \# k(S^2 \times S^2)$ ($\#$ is the connected sum operation).

It follows almost immediately from this result that if V is a simply-connected compact 4-manifold, then there exists an integer $k \geq 0$ such that $V \# (k+1)P \# kQ$ is diffeomorphic to $\ell P \# mQ$ for some $\ell, m \geq 0$, where P is $\mathbb{C}P^2$ with its usual orientation and Q is $\mathbb{C}P^2$ with orientation opposite to the usual.

After the proof of his theorem Wall writes the following ([1], p. 147): "We remark that our result is a pure existence theorem; we have obtained, even in principle, no bound whatever on the integer k".

As it was remarked in [2], the operation $V \# P$ (resp. $(V \# Q)$ where V is an oriented 4-manifold could be considered as performing of certain blowing-up of some point on V. We call this blowing-up $\bar{\sigma}$-process (resp. σ-process).

We say that an oriented compact simply-connected 4-manifold W is completely decomposable (resp. almost completely decomposable) if W (resp. $W \# P$) is diffeomorphic to $\ell P \# mQ$ for some $\ell, m \geq 0$.

Let V be an oriented compact simply-connected 4-manifold. For $(k_1,k_2) \in \mathbb{Z}\times\mathbb{Z}$, $k_1 \geq 0$, $k_2 \geq 0$, let $V(k_1,k_2)$ be a 4-manifold obtained from V by k_1 $\bar\sigma$-processes and k_2 σ-processes. Denote by $(V) = \{(k_1,k_2) \in \mathbb{Z}\times\mathbb{Z} \mid k_1 \geq 0,\ k_2 \geq 0,\ V(k_1,k_2)$ is completely decomposable$\}$. It follows from the theorem of Wall that $(V) \neq \emptyset$. An important geometrical problem is to define minimal elements of (V) (in any natural sense). A certain step for solving this problem could be the construction of some elements of (V) in explicit form, say in terms of the 2-dimensional Betti number and of the signature of V.

In the present work we show that such a construction is possible when V admits a complex structure. The main result is the following:

<u>Theorem A</u>. Let V be a compact simply-connected 4-manifold which admits a complex structure. Take an orientation on V corresponding to a certain complex structure on it. Let $K(x), L(x)$ be cubic polynomials defined as follows:

$$K(x) = \tilde{K}(g(5x+4)) - x, \qquad L(x) = \tilde{L}(g(5x+4)), \qquad \text{where}$$

$$\tilde{K}(t) = \frac{t}{3}(t^2-6t+11), \qquad \tilde{L}(t) = \frac{t-1}{3}(2t^2-4t+3)$$

$$(K(x) = 30375x^3 + 68850x^2 + 52004x + 13092,$$

$$L(x) = 60750x^3 + 141750x^2 + 110265x + 28595).$$

Denote by b_+ (corresp. b_-) the number of positive (corresp. negative) squares in the intersection form of V and let
$k_1' = K(b_+)$, $k_2' = \max(0, L(b_+)-b_-)$.

Then the pair $(k_1', k_2') \in$ (V).

From the Kodaira classification of compact complex surfaces it follows that if V is a simply-connected compact complex surface, then there exists a non-singular projective-algebraic complex surface $\tilde{V}$ such that $\tilde{V}$ is diffeomorphic to V and one of the following three possibilities holds: (a) $\tilde{V}$ is rational; (b) $\tilde{V}$ is elliptic; (c) $\tilde{V}$ is of general type. In the case (a) our theorem is evident. In the case (b) we prove a much stronger result:

<u>Theorem B</u> (see Theorem 12, §4, Part II). Any simply-connected elliptic surface V is almost completely decomposable.
(That is, $(1,0) \in$ (V)).

In the case (c) (see Theorem 5, §4, Part I) we use Bombieri's results on pluricanonical embeddings ([3]), results of [2] on the topological structure of non-singular hypersurfaces in $\mathbb{C}P^3$ and the following:

<u>Theorem C</u> (see Theorem 4, §3, Part I). Let V_n be a projective algebraic surface of degree n embedded in $\mathbb{C}P^N$, $N \geq 5$, such that V_n is not contained in a proper projective subspace of $\mathbb{C}P^N$.

Suppose that V_n is non-singular or has as singularities only rational double-points. Let $h: \tilde{V}_n \longrightarrow V$ be a minimal desingularization of V_n (that is, $\tilde{V}_n$ has no exceptional curve of the first kind s such that h(s) is a point on V_n). Denote by X_n the diffeomorphic type of a non-singular hypersurface of degree n in $\mathbb{C}P^3$.

Suppose $\pi_1(\tilde{V}_n) = 0$. Then

i) $b_+(\tilde{V}_n) < b_+(X_n)$, $b_-(\tilde{V}_n) < b_-(X_n)$,

ii) $\tilde{V}_n \# [b_+(X_n)-b_+(\tilde{V}_n)+1]P \# [b_-(X_n)-b_-(\tilde{V}_n)]Q$

is diffeomorphic to $X_n \# P$.

Note that Theorem B together with results of [2],[4],[5] shows that all big explicit classes of simply-connected algebraic surfaces considered until now have the property that their elements are almost completely decomposable 4-manifolds. That is, the "theoretical" Theorem A gives much weaker results than our "empirical knowledge". The interesting question is, how far we can move with such "empirical achievements" in more general classes of simply-connected algebraic surfaces.

I prepared this work during my visits to IHES, Bures-sur-Yvette, France, and Sonderforschungsbereich, Bonn, West Germany, in the spring of 1976. The excellent conditions which I found in these Institutes were very important (and necessary) for the appearance of this work. I am very grateful to both of these Institutes.

The Appendix to Part I is essentially based on the advice of D. Mumford. D. Mumford told me about Severi's theorem and explained its use for the proof of part 3) of Theorem 3, §3, Part I. The proof of part 4) of this theorem is also due to D. Mumford.

The idea to use in the proof of Lemma 4, §1, Part II, a non-ramified covering is due to P. Deligne.

W. Neumann and D. Husemoller read the manuscript before it was typed and made many useful remarks.

I would like to express here my deep thanks to all of them.

Part I

Topology of simply-connected algebraic surfaces of given degree n

§1. A topological comparison theorem for fibers of holomorphic functions on complex threefolds.

Lemma 1. Let U be an open subset in $\mathbb{C}^3$, $f: U \longrightarrow \Delta$ be a holomorphic function where Δ is the open unit disk in $\mathbb{C}$, such that f has only one critical point c on U, $f(c) = 0$ ($\in\Delta$) and c is a rational double-point on $f^{-1}(0)$. Let B_ϵ be a closed c-ball in U of radius ϵ and with the center c, $\mathcal{D}_{\epsilon'}$ be a closed 2-disk in Δ of radius ϵ' with the center $0 \in \Delta$ and such that $f^{-1}(\mathcal{D}_{\epsilon'})$ is transversal to $S_\epsilon = \partial B_\epsilon$ (and therefore

$$f(f^{-1}(\mathcal{D}_{\epsilon'}) \cap S_\epsilon) = \mathcal{D}_{\epsilon'}).$$

Denote

$$f_S = f|_{f^{-1}(\mathcal{D}_{\epsilon'})\cap S_\epsilon} : f^{-1}(\mathcal{D}_{\epsilon'}) \cap S_\epsilon \longrightarrow \mathcal{D}_{\epsilon'} .$$

It is clear that $f_S: f^{-1}(\mathcal{D}_{\epsilon'}) \cap S_\epsilon \longrightarrow \mathcal{D}_{\epsilon'}$ is a fibre bundle; let $\psi: f^{-1}(\mathcal{D}_{\epsilon'}) \cap S_\epsilon \longrightarrow \mathcal{D}_{\epsilon'} \times f_S^{-1}(0)$ be some trivialization of it.

Let $U_\tau = f^{-1}(\tau)$, $\tau \in \mathbb{C}$, $h: \tilde{U}_0 \longrightarrow U_0$ be a minimal desingularization of U_0 (that is, $\tilde{U}_0$ does not have an exceptional curve of the first kind s such that h(s) is a point on U_0),

$\tau' \in \mathcal{D}_{\epsilon'}$. Then there exists a diffeomorphism

$$\alpha\colon h^{-1}(U_o \cap B_\epsilon) \longrightarrow U_{\tau'} \cap B_\epsilon$$

such that

$$(\alpha|_{h^{-1}(U_o \cap S_\epsilon)}) \cdot (h^{-1}|_{U_o \cap S_\epsilon})\colon U_o \cap S_\epsilon \longrightarrow U_{\tau'} \cap S_\epsilon$$

coincides with the canonical diffeomorphism $f_S^{-1}(0) \longrightarrow f_S^{-1}(\tau')$ corresponding to the trivialization Φ of f_S.

<u>Proof</u>. We use the theory of simultaneous resolution for rational double-points (see [6],[7]). It follows from this theory that there exists a positive integer m and a commutative diagram of holomorphic maps

$$\begin{array}{ccc} U & \xleftarrow{\ \varphi_1\ } & \tilde{U} \\ {\scriptstyle f}\downarrow & & \downarrow{\scriptstyle \tilde{f}} \\ \Delta & \xleftarrow[\ \varphi\ ]{} & \tilde{\Delta} \end{array}$$

such that φ_1 is proper, $\tilde{\Delta} = \{\sigma \in \mathbb{C} \,\big|\, |\sigma| < 1\}$, $\varphi(\sigma) = \sigma^m$, $\tilde{U} - \varphi_1^{-1}(c) \xrightarrow{\tilde{f}} \tilde{\Delta}$ coincides with canonical projection $(U-c) \times_\Delta \tilde{\Delta} \longrightarrow \tilde{\Delta}$, $\varphi_1|_{\varphi_1^{-1}(U_o)}\colon \varphi_1^{-1}(U_o) \longrightarrow U_o$ coincides with $h\colon \tilde{U}_o \longrightarrow U_o$ and the function $\tilde{f}$ has no critical values.

Let $\widetilde{B} = \varphi_1^{-1}(B_\epsilon)$, $\widetilde{S} = \varphi_1^{-1}(S_\epsilon)$, $\widetilde{D} = \varphi^{-1}(D_{\epsilon'})$, $\widetilde{f}_S = \widetilde{f}|_{\widetilde{f}^{-1}(\widetilde{D})\cap\widetilde{S}} : \widetilde{f}^{-1}(\widetilde{D})\cap\widetilde{S} \longrightarrow \widetilde{D}$. Because $\varphi_1^{-1}(c) \cap \widetilde{S} = \emptyset$ we can identify $\widetilde{f}_S : \widetilde{f}^{-1}(\widetilde{D})\cap\widetilde{S} \longrightarrow \widetilde{D}$ with $[f^{-1}(D_{\epsilon'})\cap S_\epsilon] \times_{D_{\epsilon'}} \widetilde{D}$. Now using $\psi: f^{-1}(D_{\epsilon'})\cap S_\epsilon \longrightarrow D_{\epsilon'} \times f_S^{-1}(0)$ we obtain a trivialization of $\widetilde{f}_S$, $\widetilde{\psi}: \widetilde{f}^{-1}(\widetilde{D})\cap\widetilde{S} \longrightarrow \widetilde{D} \times \widetilde{f}_S^{-1}(0)$ corresponding to ψ. In particular, the canonical diffeomorphism $\widetilde{f}_S^{-1}(0) \longrightarrow \widetilde{f}_S^{-1}(\sigma)$, $\sigma \in \widetilde{D}$, corresponding to $\widetilde{\psi}$ coincides with $f_S^{-1}(0) \longrightarrow f_S^{-1}(\sigma^m)$ corresponding to ψ. Let

$$\widetilde{f}_B = \widetilde{f}|_{\widetilde{f}^{-1}(\widetilde{D})\cap\widetilde{B}} : \widetilde{f}^{-1}(\widetilde{D}) \cap \widetilde{B} \longrightarrow \widetilde{D} .$$

It is clear that $\widetilde{f}^{-1}(\widetilde{D})$ is transversal to $\partial\widetilde{B} = \widetilde{S}$. Since $\widetilde{f}$ h no critical points, $\widetilde{f}_B: \widetilde{f}^{-1}(\widetilde{D})\cap\widetilde{B} \longrightarrow \widetilde{D}$ is a differentiable fibre bundle and we can construct a trivialization $\Psi: \widetilde{f}^{-1}(\widetilde{D})\cap\widetilde{B} \longrightarrow \widetilde{D} \times [\widetilde{f}^{-1}(0)\cap\widetilde{B}]$ of $\widetilde{f}_B$ such that the diagram

$$\begin{array}{ccc} \widetilde{f}^{-1}(\widetilde{D})\cap\widetilde{S} & \xrightarrow{\ \widetilde{\psi}\ } & \widetilde{D} \times [\widetilde{f}_S^{-1}(0)] \\ \downarrow & & \downarrow \\ \widetilde{f}^{-1}(\widetilde{D})\cap\widetilde{B} & \xrightarrow{\ \Psi\ } & \widetilde{D} \times [\widetilde{f}^{-1}(0)\cap\widetilde{B}] \end{array}$$

is commutative.

Take $\sigma \in \widetilde{D} - 0$ with $(\sigma')^m = \tau'$ and let $\widetilde{\alpha}: \widetilde{f}_B^{-1}(0) \longrightarrow \widetilde{f}_B^{-1}(\sigma'$ a canonical diffeomorphism corresponding to Ψ,

$$\varphi_{1,\sigma',B} = \varphi_1 \big|_{\tilde{f}^{-1}(\sigma')\cap\tilde{B}} : \tilde{f}^{-1}(\sigma')\cap\tilde{B} \longrightarrow f^{-1}(\tau')\cap B_\epsilon .$$

Now it is easy to verify that we can take $\alpha = \varphi_{1,\epsilon',B} \cdot \tilde{\alpha}$.

Q.E.D.

Definition 1. Let W be a 3-dimensional complex manifold and let V be a complex subspace of W, $\dim_{\mathbb{C}} V = 2$. We say that the singular locus $\underline{S}(V)$ of V is canonical if V is reduced and for any $p \in \underline{S}(V)$ we have one of the following possibilities:

a) p is a rational double point of V;

b) p is an ordinary singular point of V, that is, there exists a complex coordinate neighborhood U_p of p in W such that V is defined in U_p by one of the following equations:

(i) $z_1z_2 = 0$,

(ii) $z_1z_2z_3 = 0$ (triplanar point),

(iii) $z_1^2-z_2z_3^2$ (pinch-point).

Theorem 1. Let W be a 3-dimensional complex manifold, $\Delta = \{t \in \mathbb{C} \,|\, |t| < 4\}$, $f: W \longrightarrow \Delta$ be a proper holomorphic map. Suppose that the singular locus $\underline{S}(V_o)$ of $V_o = f^{-1}(0)$ is canonical and let $\underline{R}$ denote the union of all rational double points of V_o, $S = \underline{S}(V_o)-\underline{R}$. Let $h_S: \tilde{S} \longrightarrow S$ be the normalization of S, $h: \tilde{V}_o \longrightarrow V_o$ be a minimal desingularization of V_o, $C = h^{-1}(S)$,

$h_C: \tilde{C} \longrightarrow C$ be the normalization of C, $\tau: TC \longrightarrow C$ be a regular neighborhood of C in $\tilde{V}_o$. Let $x_1, x_2, \cdots, x_\nu$ be all the triplanar points of V_o, $y_1, y_2, \cdots, y_\rho$ be all the pinch-points of V_o, $(\tilde{x}_1^\ell, \tilde{x}_2^\ell, \tilde{x}_3^\ell) = h_S^{-1}(x_\ell)$, $\ell = 1,2,\cdots,\nu$, $\tilde{y}_m = h^{-1}(y_m)$, $m = 1,2,\cdots,\rho$.

Consider ν copies of $\mathbb{C}P^2$ (with usual orientation), say $P_1, P_2, \cdots, P_\nu$ and let $(\xi_1^\ell : \xi_2^\ell : \xi_3^\ell)$ be the homogeneous coordinates in P_ℓ, $\ell = 1,2,\cdots,\nu$,

$$E_i^\ell = \{x \in P_\ell \mid \xi_i^\ell(x) = 0\},$$

$$s_i^\ell = \{x \in P_\ell \mid \xi_i^\ell(x) = 0,\ |\xi_{i'}^\ell| = |\xi_{i''}^\ell| \text{ where } (i',i'') = (1,2,3)-(i)\},$$

$$\ell = 1,2,\cdots,\nu,\quad i = 1,2,3 \quad \text{(see Fig. 1).}$$

Let T_i^ℓ be a tubular neighborhood of s_i^ℓ, $i = 1,2,3$, $\ell = 1,2,\cdots,\nu$. Take $T_1^\ell, T_2^\ell, T_3^\ell$ so that $T_1^\ell, T_2^\ell, T_3^\ell$ are pairwise disjoint; for any $i = 1,2,3$ $T_i^\ell \cap E_i^\ell$ is a tubular neighborhood of s_i^ℓ in E_i^ℓ and $T_i^\ell \cap E_{i'}^\ell = \emptyset$, $T_i^\ell \cap E_{i''}^\ell = \emptyset$ where $(i',i'') = (1,2,3)-(i)$.

Then there exist:

a) a complex-analytic projective line bundle $\pi: A \longrightarrow \tilde{S}$;

b) a differential embedding $i_C: \tilde{C} \longrightarrow A$ such that $\pi \cdot i_C: \tilde{C} \longrightarrow \tilde{S}$ coincides with the canonical map corresponding

Fig. 1

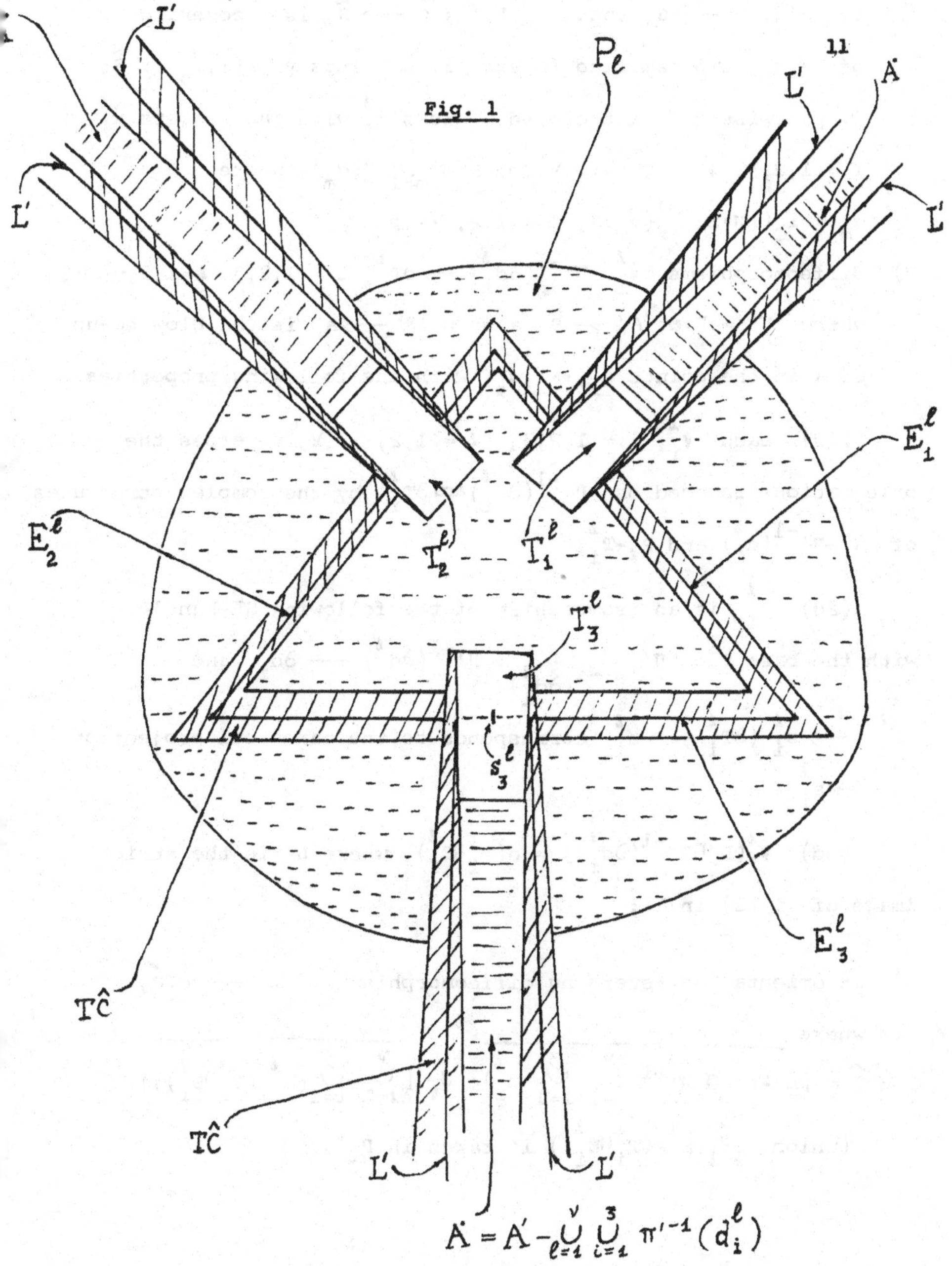
L'
L'
L'
L'
L'
L'
P_ℓ
A'
E_1^ℓ
E_2^ℓ
T_2^ℓ
T_1^ℓ
T_3^ℓ
s_3^ℓ
E_3^ℓ
TĈ
TĈ
$\dot{A} = A' - \bigcup_{\ell=1}^{\nu} \bigcup_{i=1}^{3} \pi'^{-1}(d_i^\ell)$

to $h|_C: C \longrightarrow S$, that is, $\pi \cdot i_C: \widetilde{C} \longrightarrow \widetilde{S}$ is a covering of degree two ramified in exactly ρ points $\widetilde{y}_1, \cdots, \widetilde{y}_\rho$ of $\widetilde{S}$;

c) 3ν pairwise disjoint closed 2-disks d_i^ℓ with the centers $\widetilde{x}_i^\ell$, $i = 1,2,3$, $\ell = 1,2,\ldots,\nu$, on $\widetilde{S} - \bigcup_{m=1}^{\rho} \pi(q_m)$, where $q_m = i_C((\pi \cdot i_C)^{-1}(\widetilde{y}_m))$, $m = 1,2,\cdots,\rho$;

d) diffeomorphisms $\psi_i^\ell: \pi'^{-1}(\partial d_i^\ell) \to \partial T_i^\ell$, $i = 1,2,3$, $\ell=1,2,\cdots,\nu$, where $\pi' = \pi \cdot \varphi: A' \to \widetilde{S}$ and $\varphi: A' \longrightarrow A$ is the blowing-up of A in the points $q_1, \cdots, q_\rho$, with the following properties:

(1d) each ψ_i^ℓ, $i = 1,2,3$, $\ell = 1,2,\cdots,\nu$, reverses the orientations induced on $\pi'^{-1}(\partial d_i^\ell)$ and ∂T_i^ℓ by the complex structures of $A' - \pi'^{-1}(d_i^\ell)$ and $P_\ell - T_i^\ell$;

(2d) ψ_i^ℓ is an isomorphism of the following S^2-bundles with the base S^1: $\pi'|_{\pi'^{-1}(\partial d_i^\ell)}: \pi'^{-1}(\partial d_i^\ell) \longrightarrow \partial d_i^\ell$ and $\partial T_i^\ell \longrightarrow s_i^\ell$ ($\partial T_i^\ell \longrightarrow s_i^\ell$ corresponds to the canonical projection $T_i^\ell \longrightarrow s_i^\ell$);

(3d) $\psi_i^\ell(L' \cap \pi'^{-1}(\partial d_i^\ell)) = \partial(T_i^\ell \cap E_i^\ell)$, where L' is the strict image of $i_C(\widetilde{C})$ in A';

e) an orientation reversing diffeomorphism $\eta: \partial TC \longrightarrow \partial T\widehat{C}$, where

$$\widehat{C} = \overline{[L' - L' \cap \pi'^{-1}(\bigcup_{\ell=1}^{\nu} \bigcup_{i=1}^{3} d_i^\ell)]} \cup_\psi \overline{[\bigcup_{\ell=1}^{\nu} \bigcup_{i=1}^{3} (E_i^\ell - (T_i^\ell \cap E_i^\ell))]}$$

(union $\bigcup_{i=1}^{3} (E_i^\ell - (T_i^\ell \cap E_i^\ell))$ is taken in P_ℓ),

$$\overline{\Psi}: L' \cap [\bigcup_{\ell=1}^{\nu} \bigcup_{i=1}^{3} \pi'^{-1}(\partial d_i^{\ell})] \longrightarrow \bigcup_{\ell=1}^{\nu} \bigcup_{i=1}^{3} \partial(T_i^{\ell} \cap E_i^{\ell})$$

is the diffeomorphism which is equal to $\Psi_i^{\ell}\big|_{L' \cap \pi'^{-1}(\partial d_i^{\ell})}$

on the corresponding connected components, $\hat{\tau}: T\hat{C} \longrightarrow \hat{C}$

is a regular neighborhood of $\hat{C}$ on $\hat{A}$, where

$$\hat{A} = A' - \overline{\bigcup_{\ell=1}^{\nu} \bigcup_{i=1}^{3} \pi'^{-1}(d_i^{\ell})} \cup_{\Psi} [\overline{\bigcup_{\ell=1}^{\nu}(P_{\ell} - \bigcup_{i=1}^{3} T_i^{\ell})}],$$

$\Psi: \bigcup_{\ell=1}^{\nu} \bigcup_{i=1}^{3} \pi'^{-1}(\partial d_i^{\ell}) \longrightarrow \bigcup_{\ell=1}^{\nu} \bigcup_{i=1}^{3} (\partial T_i^{\ell})$ is the diffeomorphism which is equal to Ψ_i^{ℓ} on the corresponding connected components, and the orientations of ∂TC and $\partial T\hat{C}$ correspond to the orientations of TC and $T\hat{C}$ which are defined by the complex structures of $\tilde{V}_o$ and A';

f) open subsets $U \subset C$, $\hat{U} \subset \hat{C}$ and a diffeomorphism $\eta_o: U \to \hat{U}$ such that (1) $\eta(\underline{\tau}^{-1}(U)) = \underline{\hat{\tau}}^{-1}(\hat{U})$ where $\underline{\tau} = \tau|_{\partial TC}$, $\underline{\hat{\tau}} = \hat{\tau}|_{\partial T\hat{C}}$, and (2) the diagram

$$\begin{array}{ccc} \underline{\tau}^{-1}(U) & \xrightarrow{\ \eta|\underline{\tau}^{-1}(U)\ } & \underline{\hat{\tau}}^{-1}(\hat{U}) \\ \downarrow{\scriptstyle \underline{\tau}} & & \downarrow{\scriptstyle \underline{\tau}} \\ U & \xrightarrow{\ \eta_o\ } & \hat{U} \end{array}$$

is commutative;

g) diffeomorphisms $u_t: \overline{(\widetilde{V}_o - TC)} \cup_\eta \overline{(\widehat{A} - T\widehat{C})} \longrightarrow V_t$, where $t \in \Delta - (o)$, $V_t = f^{-1}(t)$.

<u>Proof</u>. Let B_ℓ, B'_m, $\ell = 1,2,\cdots,\nu$, $m = 1,2,\cdots,\rho$, be open small pairwise disjoint coordinate 6-balls on W such that for any $\ell = 1,2,\cdots,\nu$ (corresp. $m = 1,2,\cdots,\rho$) the center of B_ℓ (corresp. B'_m) is x_ℓ (corresp. y_m) and $V_o \cap B_\ell$ (corresp. $V_o \cap B'_m$) is defined in B_ℓ (corresp. B'_m) by the local equation $z_1^{(\ell)} z_2^{(\ell)} z_3^{(\ell)} = 0$ (corresp. $(z_2'^{(m)})^2 - z_3'^{(m)}(z_1'^{(m)})^2$) where $z_1^{(\ell)}, z_2^{(\ell)}, z_3^{(\ell)}$ (corresp. $z_1'^{(m)}, z_2'^{(m)}, z_3'^{(m)}$) are complex coordinates in B_ℓ (corresp. B'_m).

Let $\widetilde{\tau}: \widetilde{T} \longrightarrow S$ be a regular neighborhood of S in W such that for $\ell = 1,2,\cdots,\nu$ and $m = 1,2,\cdots,\rho$

$$\partial\widetilde{T} \cap \partial\overline{B}_\ell \neq \emptyset, \quad \partial\widetilde{T} \cap \partial\overline{B}'_m \neq \emptyset,$$

$\partial\widetilde{T}$ is transversal to $\partial\overline{B}_\ell$ and to $\partial\overline{B}'_m$ and if

$$S' = S - [S \cap ((\bigcup_{\ell=1}^{\nu} B_\ell) \cup (\bigcup_{m=1}^{\rho} B'_m))] ,$$

$$\widetilde{T}' = \widetilde{T} - [\widetilde{T} \cap ((\bigcup_{\ell=1}^{\nu} B_\ell) \cup (\bigcup_{m=1}^{\rho} B'_m))] ,$$

then $\widetilde{\tau}(\widetilde{T}') = S'$ and $\widetilde{\tau}\big|_{\widetilde{T}'}: \widetilde{T}' \longrightarrow S'$ is a 4-disk bundle.

Let $\Delta' = \{t \in \mathbb{C} \,\big|\, |t| < \frac{3}{2}\}$. Changing if necessary the coordinate t we can assume that $\forall t \in \Delta'$, V_t is transversal to $\partial\widetilde{T}$.

Consider the ball B_1. We can assume that complex coordinates z_1,z_2,z_3 in B_1 are chosen so that $f|_{B_1} = z_1z_2z_3$. [Recall that $z_1^{(1)}z_2^{(1)}z_3^{(1)} = 0$ is a local equation of $V_o \cap B_1$ in B_1. Since $f^{-1}(0)$ has no multiple components, $f_1 = \frac{f}{z_1^{(1)}z_2^{(1)}z_3^{(1)}}$ is an invertible holomorphic function in some neighborhood of x_1. We take $z_1 = f_1 \cdot z_1^{(1)}$, $z_2 = z_2^{(1)}$, $z_3 = z_3^{(1)}$ and choose B_1 and $\tilde{T}$ smaller]. Changing, if necessary, z_1,z_2,z_3,t (t is the complex coordinate in Δ) we can identify B_1 with the open ball $\{(z_1,z_2,z_3) \in \mathbb{C}^3 \mid \sum_{i=1}^{3} |z_i|^2 < 3\}$ in $\mathbb{C}^3$ and $f|_{B_1}$ with the function $z_1z_2z_3$.

Let $Z = V_1 \cap B_1 = \{(z_1,z_2,z_3) \in \mathbb{C}^3 \mid z_1z_2z_3 = 1, \ \sum_{i=1}^{3} |z_i|^2 < 3\}$, δ be a small positive number,

$$\Pi_o = \{(z_1,z_2,z_3) \in \mathbb{C}^3 \mid |z_i| \leq 1+\delta, \ i = 1,2,3\}, \text{ for } j = 1,2,3$$

$$\Pi_j = \{(z_1,z_2,z_3) \in \mathbb{C}^3 \mid 1+\delta \leq |z_j| \leq 2, \ |z_{j'}| \leq 1, \ |z_{j''}| \leq 1, \text{ where } (j',j'') = (1,2,3)-(j)\},$$

$$\tilde{M}_j = \Pi_j \cap Z, \quad \tilde{M}_o = \Pi_o \cap Z, \quad \tilde{M} = \bigcup_{j=0}^{3} \tilde{M}_j.$$

Consider a 3-dimensional real euclidean space $\mathbb{R}^3$ with coordinates r_1,r_2,r_3 (see Fig. 2) and let $\mathbb{R}^3_+ = \{(r_1,r_2,r_3) \in \mathbb{R}^3 \mid r_i \geq 0, \ i=1,2,3\}$

Fig. 2

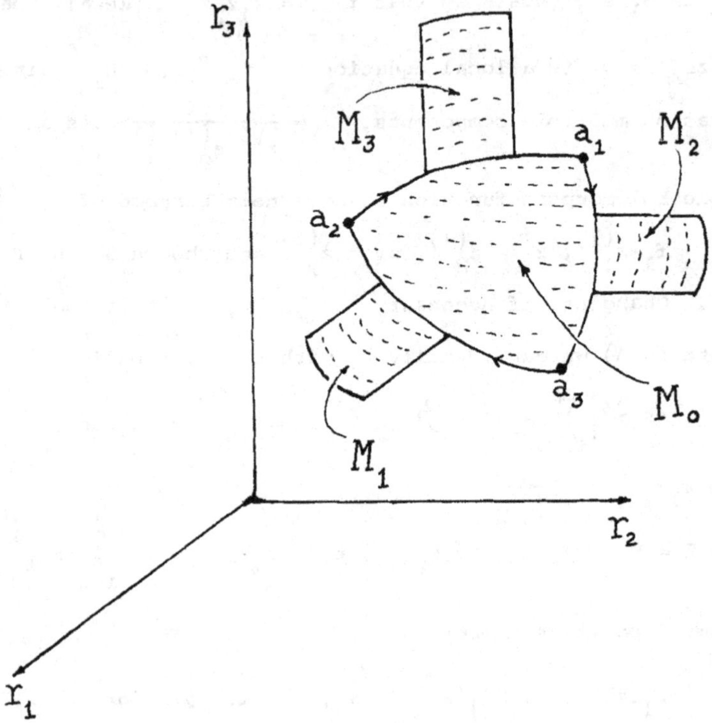

$M_o = \{(r_1,r_2,r_3) \in \mathbb{R}^3_+ \mid r_1r_2r_3 = 1,\ r_i \leq 1+\delta,\ i = 1,2,3\}$, for $j=1,2,3$

$M_j = \{(r_1,r_2,r_3) \in \mathbb{R}^3_+ \mid r_1r_2r_3 = 1,\ 1+\delta \leq r_j \leq 2,\ r_{j'} \leq 1,\ r_{j''} \leq 1,$

where $(j',j'') = (1,2,3)-(j)\}$, $M = \bigcup_{j=o}^{3} M_j$.

Identify $U(1)$ with $\{g \in \mathbb{C} \mid |g| = 1\}$ and let $T^3_Z = U(1)\times U(1)\times U(1)$ with coordinates g_1,g_2,g_3, and T be the subgroup of T^3_Z defined by the equation: $g_1g_2g_3 = 1$. Writing $z_i = r_ig_i$ where $r_i \in \mathbb{R}$, $r_i > 0$, $g_i \in \mathbb{C}$, $|g_i| = 1$, we identify $\widetilde{M}_o$ with $M_o\times T$, $\widetilde{M}_j$ with $M_j\times T$ and $\widetilde{M}$ with $M\times T$.

It is clear that M_o is diffeomorphic to the triangle in $\mathbb{R}^3$ with the vertexes: $a_1 = (\frac{1}{(1+\delta)^2}, 1+\delta, 1+\delta)$, $a_2 = (1+\delta, \frac{1}{(1+\delta)^2}, 1+\delta)$, $a_3 = (1+\delta, 1+\delta, \frac{1}{(1+\delta)^2})$. Let us take on M the orientation, which on M_o coincides with the orientation corresponding to the order (a_1,a_3,a_2) of the vertexes a_1,a_2,a_3.

Note that the map $T \longrightarrow U(1)\times U(1)$ given by $(g_1,g_2,g_3) \longrightarrow (g_1,g_2)$ is an isomorphism (of groups). Using this isomorphism and the canonical orientation on $U(1)\times U(1)$ (corresponding to the given order of factors and to the positive rotation on $U(1)$) we fix an orientation of T. It is easy to verify that the orientation on $M\times T$ which we now obtain coincides with the orientation of $M\times T = \widetilde{M}$ corresponding to the complex structure on $\widetilde{M}$.

Consider a new copy of $\mathbb{R}^3$ with coordinates ρ_1, ρ_2, ρ_3 (see Fig. 3) and let

$$\underline{\Delta} = \{(\rho_1,\rho_2,\rho_3) \mid \rho_1+\rho_2+\rho_3 = 3, \quad \rho_i \geq 0, \ i = 1,2,3\},$$

R_o be the triangle in $\underline{\Delta}$ with the vertexes

$$b_1 = (2, \tfrac{1}{2}, \tfrac{1}{2}), \ b_2 = (\tfrac{1}{2}, 2, \tfrac{1}{2}), \ b_3 = (\tfrac{1}{2}, \tfrac{1}{2}, 2),$$

for $j = 1,2,3$

R_j be a polygon in Δ which is formed by the following four straight lines:

1) $\rho_j = \frac{1}{2}$; 2) $\rho_j = \frac{1}{3}$; 3) $\rho_{j'} = 2\rho_{j''}$; 4) $\rho_{j''} = 2\rho_{j'}$,

$$(j',j'') = (1,2,3)-(j),$$

$$R = \bigcup_{j=0}^{3} R_j \ .$$

Let T_p^3 be a new copy of $U(1)\times U(1)\times U(1)$ with coordinates h_1,h_2,h_3, H a subgroup in T_p^3, defined by the equations $h_1=h_2=h_3$. Denote $T_p = T_p^3/H$.

Consider $P = \mathbb{C}P^2$ with homogeneous coordinates ξ_1,ξ_2,ξ_3 chosen so that $|\xi_1| + |\xi_2| + |\xi_3| = 3$. Let for $j = 0,1,2,3$

$$\tilde{R}_j = \{(\xi_1,\xi_2,\xi_3) \in P \mid (|\xi_1|,|\xi_2|,|\xi_3|) \in R_j\}$$

$$\tilde{R} = \bigcup_{j=0}^{3} \tilde{R}_j .$$

Fig. 3

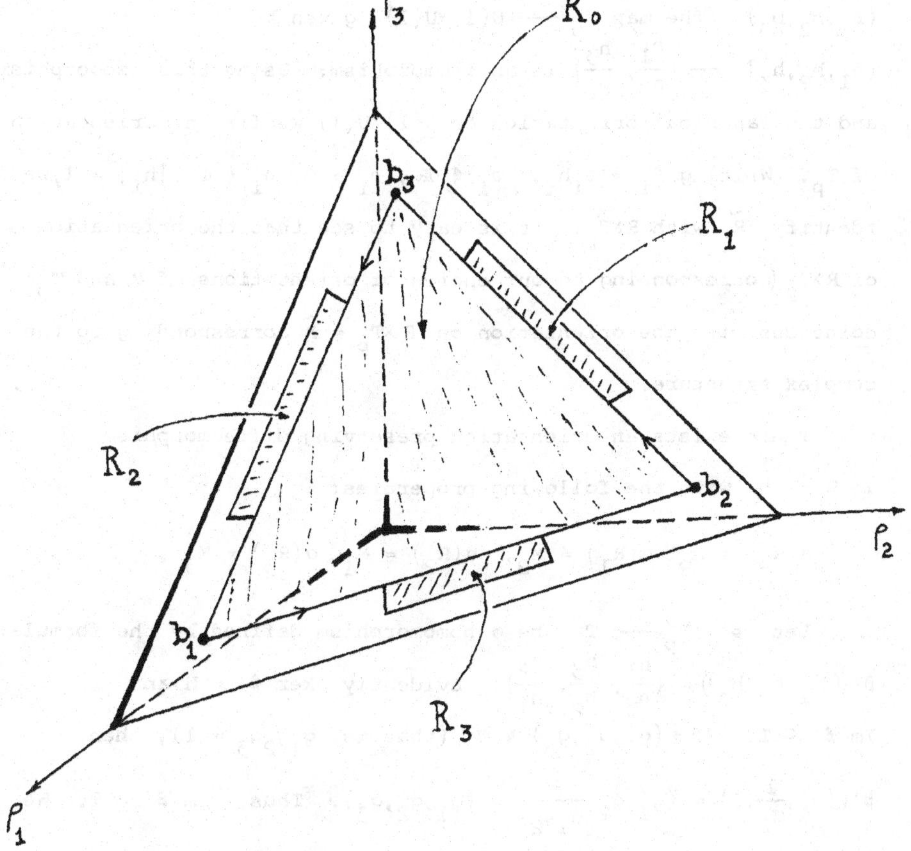

Take on R the orientation which on R_o coincides with the orientation corresponding to the following order of vertexes: (b_1,b_2,b_3). The map $T_p \longrightarrow U(1)\times U(1)$, given by $[h_1,h_2,h_3] \longrightarrow (\frac{h_1}{h_3}, \frac{h_3}{h_2})$ is an isomorphism. Using this isomorphism and the canonical orientation on $U(1)\times U(1)$ we fix an orientation of T_p. Writing $\xi_i = \rho_i h_i$, $\rho_i \in \mathbb{R}$, $\rho_i > 0$, $h_i \in \mathbb{C}$, $|h_i| = 1$, we identify $\tilde{R}$ with $R\times T_p$. It is easy to see that the orientation of $R\times T_p$ (corresponding to our choice of orientations of R and T_p) coincides with the orientation on $R\times T_p = \tilde{R}$ corresponding to the complex structure.

There exists an orientation preserving diffeomorphism $\alpha: R \longrightarrow M$ with the following properties:

$$\alpha(R_o) = M_o,\ \alpha(R_1) = M_2,\ \alpha(R_2) = M_1,\ \alpha(R_3) = M_3 .$$

Let $\beta': T_p^3 \longrightarrow T^3$ be a homomorphism defined by the formula: $\beta'((h_1,h_2,h_3)) = (\frac{h_1}{h_3}, \frac{h_3}{h_2}, \frac{h_2}{h_1})$. Evidently $\ker \beta' = H$ and $\operatorname{Im} \beta' \subset T$. If $(g_1,g_2,g_3) \in T$ (that is, $g_1g_2g_3 = 1$), then $\beta'((g_1,\frac{1}{g_2},1)) = (g_1,g_2,\frac{1}{g_1g_2}) = (g_1,g_2,g_3)$. Thus $\operatorname{Im} \beta' = T$. We see that β' defines an isomorphism $\beta: T_p \longrightarrow T$. Taking $\tilde{\alpha} = \alpha\times\beta: \tilde{R}(= R\times T_p) \longrightarrow \tilde{M}(= M\times T)$ we can check that $\tilde{\alpha}$ preserves orientation.

Let (i,j,k) be one of the triples $(1,2,3),(2,3,1),(3,1,2)$. Identify $\mathbb{C}^3$ with $\mathbb{C}^1 \times \mathbb{C}^2$ by the rule: $(z_1,z_2,z_3) \to (z_i,(z_j,z_k))$. Taking the projective closure of $\mathbb{C}^2$ we embed $\mathbb{C}^3$ in $\mathbb{C}^1 \times \mathbb{C}P^2$. Let homogeneous coordinates in $\mathbb{C}P^2$ be η_o, η_j, η_k and let $z_j = \frac{\eta_j}{\eta_o}$, $z_k = \frac{\eta_k}{\eta_o}$. Denote

$$M_i' = \{(z_1,z_2,z_3) \in \mathbb{C}^3,\ z_1 z_2 z_3 = 1,\ 2 \leq |z_i| \leq 2{,}5,\ |z_j| \leq 1, |z_k| \leq 1\},$$

$$K_{1,i} = \{(z_1,z_2,z_3) \in \mathbb{C}^3,\ 2 \leq |z_i| \leq 2{,}5,\ z_j = 0,\ z_k = 0\}$$

and let $\pi_i : M_i' \longrightarrow K_{1,i}$ be defined by

$$z_i(\pi_i(x)) = z_i(x),\ z_j(\pi_i(x)) = 0,\ z_k(\pi_i(x)) = 0,\ x \in M_i'.$$

π_i defines on M_i' the structure of a fiber bundle with the base $K_{1,i}$ and the fibers

$$\pi_i^{-1}(y) = \{(z_1,z_2,z_3) \in \mathbb{C}^3 \,\big|\, z_j z_k = \frac{1}{z_i(y)},\ |z_j| \leq 1,\ |z_k| \leq 1\}.$$

The embedding $\mathbb{C}^3 \longrightarrow \mathbb{C}^1 \times \mathbb{C}P^2$ which we have constructed above gives us a compactification $\overline{\pi}_i : \overline{M}_i \longrightarrow K_{1,i}$ of $\pi_i : M_i' \longrightarrow K_{1,i}$ where $\forall\, y \in K_{1,i}$, $\overline{\pi}_i^{-1}(y)$ is a non-singular rational curve in $\mathbb{C}P^2$ defined by the equation $\eta_j \eta_k = \frac{\eta_o^2}{z_i(y)}$. Let C_{ji} (corresp. C_{ki}) be a subset of $\overline{M}_i$ defined by $\eta_o = \eta_k = 0$ (corresp. $\eta_o = \eta_j = 0$), and $T_{ji} \subset \overline{M}_i$ (corresp. $T_{ki} \subset \overline{M}_i$) be defined by $|\eta_k| \leq \frac{1}{z_i}|\eta_o|$ (corresp. $|\eta_j| \leq \frac{1}{z_i}|\eta_o|$),

$Y_i = \{(z_1,z_2,z_3) \in \mathbb{C}^3,\ |z_i| = 2,\ z_j = 0,\ z_k = 0\}$,

$\Gamma_i = \overline{\pi}^{-1}(Y_i)$, $\Gamma'_{ji} = \Gamma_i \cap T_{ji}$, $\Gamma'_{ki} = \Gamma_i \cap T_{ki}$, $H_{ji} = \partial\Gamma'_{ji}$, $H_{ki} = \partial\Gamma'_{ki}$,

$L_{mn} = \{(\xi_1:\xi_2:\xi_3) \in P \,\big|\, |\xi_m| = \frac{1}{2}|\xi_n|,\ |\xi_{m'}| \leq \frac{1}{3},\ m' = (1,2,3)-(m,n)\}$,

where $m = 1,2,3$, $n = 1,2,3$,

$L^{\bullet}_{mn} = \partial L_{mn}$.

Let $\mathcal{J} = \{1,2,3\}$ and

$$\varphi\colon [(\mathcal{J}\times\mathcal{J})\text{-diagonal}] \longrightarrow [(\mathcal{J}\times\mathcal{J})\text{-diagonal}]$$

be a 1-1 map defined as follows

$$\varphi((1,2)) = (2,3),\quad \varphi((1,3)) = (2,1),\quad \varphi((2,3)) = (1,2),$$

$$\varphi((2,1)) = (1,3),\quad \varphi((3,1)) = (3,1),\quad \varphi((3,2)) = (3,2).$$

It is easy to see that $\overline{\pi}_i\big|_{T_{ji}}$ (corresp. $\overline{\pi}_i\big|_{T_{ki}}$) defines on T_{ji} (corresp. T_{ki}) the structure of a 2-disks fiber bundle over C_{ji} (corresp. C_{ki}). Thus we can consider T_{ji} (corresp. T_{ki}) with this projection as a tubular neighborhood of C_{ji} (corresp. C_{ki}) in $\overline{M}_i$.

Because $H_{ji}, H_{ki} \subset \widetilde{M}$ we can construct a new space

$$\overline{M} = \widetilde{M} \cup_{H_{21}} \Gamma'_{21} \cup_{H_{31}} \Gamma'_{31} \cup_{H_{13}} \Gamma'_{13} \cup_{H_{23}} \Gamma'_{23} \cup_{H_{32}} \Gamma'_{32} \cup_{H_{12}} \Gamma'_{12}\,.$$

A direct verification shows that $\widetilde{\alpha}(L^{\bullet}_{mn}) = H_{\varphi(m,n)}$. We obtain diffeomorphisms $\mu_{mn}\colon L^{\bullet}_{mn} \longrightarrow H_{\varphi(m,n)}$ where $\mu_{mn} = \widetilde{\alpha}\big|_{L^{\bullet}_{mn}}$.

Note that $\dot{L}_{mn}$ (corresp. L_{mn}) is a circle (corresp. 2-disk) fibre bundle over a circle in P defined by $\xi_{m'} = 0$, $|\xi_m| = \frac{1}{2}|\xi_n|$. In the notations used above a fibre of the fibration $\dot{L}_{mn}$ is given by:

$$\frac{h_m}{h_n} = b = \text{const} \in U(1), \quad |\xi_{m'}| = \frac{1}{3}, \quad |\xi_m| = \frac{1}{2}|\xi_n|.$$

Because $\beta(h_1,h_2,h_3) = (\frac{h_1}{h_3}, \frac{h_3}{h_2}, \frac{h_2}{h_1})$ we have

$$\beta(\{\frac{h_2}{h_1} = b\}) = \{g_3 = b\}$$

$$\beta(\{\frac{h_1}{h_2} = b\}) = \{g_3 = \frac{1}{b}\}$$

$$\beta(\{\frac{h_3}{h_2} = b\}) = \{g_2 = b\}$$

$$\beta(\{\frac{h_2}{h_3} = b\}) = \{g_2 = \frac{1}{b}\}$$

$$\beta(\{\frac{h_1}{h_3} = b\}) = \{g_1 = b\}$$

$$\beta(\{\frac{h_3}{h_1} = b\}) = \{g_1 = \frac{1}{b}\}.$$

Considering H_{ji}, H_{ki} as circle-bundles over Y_i (with projection $\overline{\pi}_i$) we see that the fibers of H_{ji}, H_{ki} are given by the condition $g_i = \text{const}$. Hence μ_{mn} is a diffeomorphism of circle-bundles

$\dot{L}_{mn}$ and $H_{\varphi(m,n)}$. Thus we can extend μ_{mn} to a diffeomorphism of corresponding 2-disk bundles $\bar{\mu}_{mn}: L_{mn} \longrightarrow \Gamma'_{\varphi(m,n)}$, which transforms the centers of fibers to the centers of fibers. Considering $\bar{\mu}^{-1}_{mn}$ and $\tilde{\alpha}^{-1}$ together we obtain an embedding

$$\lambda: \bar{M} \longrightarrow P .$$

Let

$$s_{m'} = \{(\xi_o:\xi_1:\xi_2) \in P \,|\, \xi_{m'} = 0, \quad |\xi_m| = |\xi_n|\},$$

$$L'_{mn} = \{(\xi_1:\xi_2:\xi_3) \in P, \quad |\xi_{m'}| = \tfrac{1}{3}, \tfrac{1}{2}\xi_n \leq \xi_m \leq 2\xi_n\},$$

$$\mathcal{H}_{m'} = L_{mn} \cup L'_{mn} \cup L_{nm} \text{ (union in P).}$$

We can consider $\mathcal{H}_{m'}$ as the boundary of a tubular neighborhood $J_{m'}$ of $s_{m'}$ in P. A fiber of the corresponding 2-sphere fibration is defined as follows:

If $b = \frac{h_m}{h_n}$ (we can identify b with a point of $s_{m'}$) then the corresponding fiber F_b is equal to the union of the fiber $L_{mn}(b)$ of L_{mn} over b, the subset of P, defined by $\arg \frac{\xi_m}{\xi_n} = \arg b$, $|\xi_{m'}| = \frac{1}{3}$, $\frac{1}{2}|\xi_n| \leq |\xi_m| \leq 2|\xi_n|$, and the fiber $L_{nm}(b)$ of L_{nm} over b.

Now let (m,n) be one of the pairs (1,3),(2,1),(3,2). Take on $s_{m'}$ the orientation corresponding to the positive rotation of $\frac{h_m}{h_n}$. Then the orientation on $\mathcal{H}_{m'} = \partial \mathcal{T}_{m'}$ defined by the complex structure on $\mathcal{T}_{m'}$ induces on every F_b an orientation which on $L_{mn}(b)$ is opposite to the orientation corresponding to the complex structure of the complex projective line $\xi_m = \frac{1}{2}b\xi_n$. But that means that the orientation on $\mathcal{H}_{m'}$, considered as $\partial(P-\mathcal{T}_{m'})$, defines on F_b an orientation which coincides on $L_{mn}(b)$ with the orientation given by the complex structure of the line $\{\xi_m = \frac{1}{2}b\xi_n\}$. This "complex" orientation on $L_{mn}(b)$ defines on $\dot{L}_{mn}(b) = \partial L_{mn}(b)$ an orientation which corresponds to the positive rotation of $\frac{h_{m'}}{h_n}$.

Since $\beta((h_1,h_2,h_3)) = (\frac{h_1}{h_3}, \frac{h_3}{h_2}, \frac{h_2}{h_1})$ we have that μ_{mn} defines on $\mu_{13}(\dot{L}_{13}(b))$ (corresp. $\mu_{21}(\dot{L}_{21}(b))$, corresp. $\mu_{32}(\dot{L}_{32}(b))$) the same orientation as the positive rotation of $\frac{1}{g_2}$ (corresp. $\frac{1}{g_1}$, corresp. $\frac{1}{g_3}$) does.

Let $\psi: \mathcal{J} \longrightarrow \mathcal{J}$ be given by $\psi(1) = 2$, $\psi(2) = 1$, $\psi(3) = 3$. A direct verification shows that $\lambda(\Gamma_i) = \mathcal{H}_{\psi(i)}$. Let $\lambda_i = \lambda|_{\Gamma_i} : \Gamma_i \longrightarrow \mathcal{H}_{\psi(i)}$. Let us fix on $\mathcal{H}_{m'}$ the orientation which is given by considering $\mathcal{H}_{m'}$ as $\partial(P-\mathcal{T}_{m'})$ and by the complex structure of $P-\mathcal{T}_{m'}$. We see now that λ_i induces on the fibers of $\Gamma_i \longrightarrow Y_i$ an orientation which coincides with the orientation given by the complex structure of these fibers. We can check also that if $\lambda_i^o: Y_i \longrightarrow S_{\psi(i)}$ is the diffeomorphism of bases corresponding canonically to λ_i then λ_i^o induces on γ_i the positive orientation (we consider γ_i as a circle in the z_i-axis in $\mathbb{C}^3$). Take on Γ_i the orientation which is given by $\Gamma_i \subset \partial \bar{M}_i$ and by the complex structure of $\bar{M}_i - \partial \bar{M}_i$. We obtain that λ_i is an orientation reversing diffeomorphism. Define $\lambda': \bigcup_{i=1}^{3} \Gamma_i \to \bigcup_{m'=1}^{3} \mathcal{H}_{m'}$ by $\lambda'|_{\Gamma_i} = \lambda_i$ and let $\hat{P} = (\bigcup_{i=1}^{3} \bar{M}_i) \cup_{\lambda'} [P - \bigcup_{m'=1}^{3} \mathcal{J}_{m'}]$. Using the embeddings $M_i' \longrightarrow \bar{M}_i$, $\lambda: \bar{M} \longrightarrow P$ which we constructed above we obtain an embedding $\hat{\lambda}: \hat{Z} \longrightarrow \hat{P}$, where $\hat{Z} = Z \cap \hat{T}^{(1)}$,

$$\hat{T}^{(1)} = (\bigcup_{j=0}^{3} \Pi_j) \cup (\bigcup_{j=1}^{3} \Pi_j'),$$

$$\Pi_j' = \{(z_1,z_2,z_3) \in \mathbb{C}^3 \,|\, 2 \le |z_j| \le 2{,}5, \quad |z_{j'}| \le 1, \quad |z_{j''}| \le 1,$$
$$(j',j'') = (1,2,3)-(j)\}.$$

Note that we can consider $\hat{T}^{(1)}$ as a part of the regular neighborhood $\tilde{T}$ of S in W and $\hat{P}-\hat{\lambda}(\hat{Z})$ as a regular neighborhood in $\hat{P}$ of the subcomplex $C^{(1)}$ in $\hat{P}$ which is equal to

$$(\bigcup_{i=1}^{3} (C_{ji} \cup C_{ki})) \cup (\bigcup_{m'=1}^{3} (E_{m'} - \mathcal{J}_{m'} \cap E_{m'})),$$

where $E_{m'} = \{(\xi_1:\xi_2:\xi_3) \in P \,|\, \xi_{m'} = 0\}$, $(j,k) = (1,2,3)-(i)$.

Now we can do the same constructions for each B_ℓ, $\ell = 2,3,\cdots,\nu$, as we did for B_1. We obtain then for each $\ell = 1,2,\cdots,\nu$ a set $\hat{Z}_\ell \subset W$, a manifold with boundary $\hat{P}_\ell$, annuluses $K_{\ell,1}$, $K_{\ell,2}$, $K_{\ell,3}$ on S, fibrations $\bar{M}_{\ell,i} \longrightarrow K_{\ell,i}$, $i = 1,2,3$, cross-sections $C_{\ell,ji}$, $C_{\ell,ki}$, $(j,k) = (1,2,3)-(i)$, of these fibrations, tubular neighborhoods $T_{\ell,ji}$, $T_{\ell,jk}$ of these cross-sections and an embedding $\hat{\lambda}_\ell\colon \hat{Z}_\ell \longrightarrow \hat{P}_\ell$ which have the same properties as

$$\hat{Z},\ \hat{P},\ K_{1,1},\ K_{1,2},\ K_{1,3},\ \bar{M}_i \longrightarrow K_{1,i},\ C_{ji},\ C_{ki},\ T_{ji},\ T_{jk},\ \hat{\lambda}$$

constructed above.

We can define also $T^{(\ell)}C^{(\ell)} \subset P_\ell$, $\ell = 2,3,\cdots,\nu$, in the same way as we defined $T^{(1)}, C^{(1)}$ and assume that $\forall\ \ell = 1,2,\cdots,\nu$ $T^{(\ell)}$ is a part of the regular neighborhood $\tilde{T}$ of S in W and consider $\hat{P}_\ell - \hat{\lambda}_\ell(\hat{Z}_\ell)$ as a regular neighborhood of $C^{(\ell)}$ in $\hat{P}_\ell$.

Consider now the ball B_1'. We can assume that there are complex coordinates z_1, z_2, z_3 in B_1 such that $f|_{B_1'}$ is defined in B_1' by the equation $f = z_2^2 - z_3 z_1^2$. [Recall that $(z_2'^{(1)})^2 - z_3'^{(1)}(z_1'^{(1)})^2 = 0$ is a local equation of $V_o \cap B_1'$ in B_1'. Since $f^{-1}(o)$ has no multiple components

$$f_2 = \frac{f}{(z_2'^{(1)})^2 - z_3'^{(1)}(z_1'^{(1)})^2}$$ is an invertible holomorphic

function in some neighborhood of y_1. We take $z_1 = \sqrt{f_2}\, z_1'^{(1)}$, $z_2 = \sqrt{f_2}\, z_2'^{(1)}$, $z_3 = z_3'^{(1)}$ and choose B_1' and $\tilde{T}$ smaller]. Changing if necessary z_1, z_2, z_3, t we can identify B_1' with $\{(z_1, z_3, z_3) \in \mathbb{C}^3 \mid \sum_{i=1}^{3} |z_i|^2 < 3\}$ and $f|_{B_1}$ with the function $z_2^2 - z_3 z_1^2$.

Now let $\Pi = \{(z_1, z_2, z_3) \in \mathbb{C}^3 \mid |z_1| < 2,\ |z_2| < 2,\ |z_3| \le \frac{1}{2}\}$,

$$Z = \{(z_1, z_2, z_3) \in \mathbb{C}^3 \mid z_2^2 - z_3 z_1^2 = 1\}, \quad T = Z \cap \Pi .$$

Identifying $\mathbb{C}^3$ with $\mathbb{C}^2 \times \mathbb{C}^1$ we embed $\mathbb{C}^3$ in $\mathbb{C}P^2 \times \mathbb{C}^1$. We take $z_1 = \frac{\eta_1}{\eta_o}$, $z_2 = \frac{\eta_2}{\eta_o}$, where η_o, η_1, η_2 are homogeneous coordinates in $\mathbb{C}P^2$, and consider z_3 as a coordinate in $\mathbb{C}^1$. Let

$$\bar{Z} = \{(\eta_o : \eta_1 : \eta_2 ; z_3) \in \mathbb{C}P^2 \times \mathbb{C}^1 \mid \eta_2^2 - z_3 \eta_1^2 = \eta_o^2\},$$

$$\bar{T} = \{(\eta_o : \eta_1 : \eta_2 ; z_3) \in \mathbb{C}P^2 \times \mathbb{C}^1 \mid \eta_2^2 - z_3 \eta_1^2 = \eta_o^2,\ |z_3| \le \frac{1}{2}\},$$

$$C_Z = \{(\eta_o : \eta_1 : \eta_2 ; z_3) \in \bar{Z} \mid \eta_o = 0\}.$$

Note that $\eta_1 \neq 0$ on C_Z. Hence we can define C_Z also by the equation $\frac{\eta_o}{\eta_1} = 0$. It follows from this that the set $\mathfrak{J}_C = \{(\eta_o : \eta_1 : \eta_2 ; z_3) \in \bar{Z} \mid |\eta_o| \le \frac{1}{2}|\eta_1|\}$ is a tubular neighborhood of C_Z in $\bar{Z}$.

It is clear that $\mathfrak{J}_C \cap \bar{T} \subseteq \bar{T} - T$. Take $x \notin \bar{T} - \mathfrak{J}_C \cap \bar{T}$. We have $|\eta_o(x)| > \frac{1}{2}|\eta_1(x)|$, $|z_3(x)| \le \frac{1}{2}$. Hence $\eta_o(x) \neq 0$, $|z_1(x)| < 2$, $|z_2(x)| < \sqrt{1 + |z_3(x)(z_1(x))^2|} < \sqrt{3} < 2$.

We obtain $x \in T$. Thus $\mathfrak{J}_C \cap \overline{T} = \overline{T} - T$.

Let

$$K = \{z_3 \in \mathbb{C}^1 \mid \tfrac{1}{3} \leq |z_3| \leq \tfrac{1}{2}\},$$

$$\mathfrak{J}_K = \{x \in \mathfrak{J}_C \mid z_3(x) \in K\},$$

$\pi_K\colon \mathfrak{J}_K \longrightarrow K$ is defined by projection

$$(\eta_0 : \eta_1 : \eta_2 ; z_3) \longrightarrow (z_3).$$

Take $z_3^o \in K$ and consider $\pi_K^{-1}(z_3^o)$. We have $\eta_2^2 - z_3^o \eta_1^2 = \eta_o^2$, $|\eta_o| \leq \frac{1}{2}|\eta_1|$. It is clear that $\eta_1 \neq 0$, so denoting $u = \dfrac{\eta_o}{\eta_1}$, $v = \dfrac{\eta_2}{\eta_1}$, we obtain $v^2 - z_3^o = u^2$; $|u| \leq \frac{1}{2}$. From $v = \pm\sqrt{u^2 + z_3^o}$, $|u|^2 \leq \frac{1}{4}$; $|z_3^o| \geq \frac{1}{3}$ we see that the projection of $\pi_K^{-1}(z_3^o)$ is an unramified map. Hence $\pi_K^{-1}(z_3^o)$ is a disjoint union of two 2-disks d_+, d_- where

$$d_+ = \{(u,v) \in \mathbb{C}^2 \mid |u| \leq \tfrac{1}{2},\ v = +\sqrt{u^2 + z_3^o}\},$$

$$d_- = \{(u,v) \in \mathbb{C}^2 \mid |u| \leq \tfrac{1}{2},\ v = -\sqrt{u^2 + z_3^o}\}.$$

It is clear that these 2-disks are transversal to C_Z and we can consider them as fibers of the tubular neighborhoods $\mathfrak{J}_C$ of C_Z corresponding to the points:

$$z_3 = z_3^o,\quad u = 0,\quad v = +\sqrt{z_3^o}$$

and

$$z_3 = z_3^o,\quad u = 0,\quad v = -\sqrt{z_3^o}\,.$$

(Here $+\sqrt{\ }$ and $-\sqrt{\ }$ mean simply some choice of branches of $\sqrt{\ }$).

Taking for each of the constructed 2-disks its center (that is, its intersection with C_Z) we obtain a map $p_K\colon \mathfrak{J}_K \longrightarrow C_K$ where $C_K = C_Z \cap \mathfrak{J}_K$. We can assume that the canonical projection $p\colon \mathfrak{J}_C \longrightarrow C_Z$ is defined so that $p|_{\mathfrak{J}_K} = p_K$.

Note that the projection $\pi_{\overline{Z}}\colon \overline{Z} \longrightarrow \mathbb{C}^1$ $(\pi_{\overline{Z}}((\eta_0 : \eta_1 : \eta_2 ; z_3)) = z_3)$ has the following property: $\pi_{\overline{Z}}^{-1}(z_3)$ is a non-singular rational curve if $z_3 \neq 0$ and a pair of transversal non-singular rational curves if $z_3 = 0$. This means that we can consider $\pi_{\overline{Z}}$ as composition of projection $\pi'\colon \mathbb{C}P^1 \times \mathbb{C}^1 \longrightarrow \mathbb{C}^1$ and blowing-up of $\mathbb{C}P^1 \times \mathbb{C}^1$ with some center $a \in \pi'^{-1}(0)$.

Denote the embedding $T \subset \overline{T}$ by $\lambda'\colon T \longrightarrow \overline{T}$. We see from $\mathfrak{J}_C \cap \overline{T} = \overline{T} - T$ that $T - \overline{T}$ is a regular neighborhood in $\overline{T}$ of a subcomplex in $\overline{T}$ which is defined by the equation $\eta_0 = 0$.

Now we can do the same constructions for each B_m', $m = 2, \rho$, as we did for B_1. We obtain then for each $m = 1, 2, \cdots, \rho$ an open 2-disk d_m on S with center y_m, a manifold $\overline{T}_m$, a map $\overline{\pi}_m\colon \overline{T}_m \longrightarrow d_m$ with $\overline{\pi}_m = \pi_m \sigma_m$ (where $\pi_m\colon \mathbb{C}P^1 \times d_m \longrightarrow d_m$ is the canonical projection and $\sigma_m\colon \overline{T}_m \longrightarrow \mathbb{C}P^1 \times d_m$ is the σ-process with some center $a_m \in \pi_m^{-1}(y_m)$), a non-singular complex curve $C_m \subset \overline{T}_m$ such that

$\bar{\pi}_m|_{C_m} : C_m \longrightarrow d_m$ is a ramified covering of degree two with unique branch point over y_m, a tubular neighborhood $p_m : \mathcal{J}_m \longrightarrow C_m$ of C_m in $\bar{T}_m$, an embedding $\lambda'_m : T_m \longrightarrow \bar{T}_m$ with $\lambda'(T_m) = \bar{T}_m - \mathcal{J}_m$, $\tilde{\tau}|_{T_m} = \bar{\pi}_m \lambda'_m$ where $T_m = \tilde{T} \cap B'_m \cap V_1$, and an annulus $K_m \subset d_m$ with center y_m such that

$$\bar{\pi}_m|_{\bar{\pi}_m^{-1}(K_m) \cap \mathcal{J}_m} = \bar{\pi}_m \cdot (p_m|_{\bar{\pi}_m^{-1}(K_m) \cap \mathcal{J}_m}) .$$

Using Lemma 1 and the assumption that $\partial\tilde{T}$ is transveral to V_t $t \in \Delta'$ we can identify $\overline{V_1 - V_1 \cap \tilde{T}}$ with $\tilde{V}_o - h^{-1}(V_o \cap \tilde{T})$. We can consider $h^{-1}(V_o \cap \tilde{T})$ as a regular neighborhood of C in $\tilde{V}_o$. Denote $TC = h^{-1}(V_o \cap \tilde{T})$ and let the corresponding projection be $\tau : TC \longrightarrow C$.

Consider the annulus $h_S^{-1}(K_{\ell,i})$ on $\tilde{S}$, $\ell = 1,2,\cdots,\nu$, $i = 1,2,3$. We can assume that $\tilde{x}_i^\ell$ is the center of $h_S^{-1}(K_{\ell,i})$. Let $\tilde{d}_{io}^\ell$ be the closed 2-disk on $\tilde{S}$ with center $\tilde{x}_i^\ell$ such that $\tilde{d}_{io}^\ell \cap h_S^{-1}(K_{\ell,i}) = \partial\tilde{d}_{io}^\ell$. Denote $d_{io}^\ell = h_S(\tilde{d}_{io}^\ell)$. Let d_{mo} be the closed 2-disk on $\tilde{S}$ with center y_m such that $d_{mo} \cap K_m = \partial d_{mo}$. Denote $S'' = S - [(\bigcup_{\ell=1}^{\nu} d_{io}^\ell) \cup (\bigcup_{m=1}^{\rho} d_{mo}]$, $C'' = h^{-1}(S'')$, $TC'' = \tau^{-1}(C'')$, $\tau'' = \tau|_{TC''} : TC'' \longrightarrow C''$. Let $q : N \longrightarrow C''$ be the normal bundle of C'' in $\tilde{V}_o$, $\hat{q} : \hat{N} \longrightarrow C''$ be the projective closure of $q : N \longrightarrow C''$. We can assume that TC'' is a subspace of N $(\subset \hat{N})$ and $\tau'' = q|_{TC''} = \hat{q}|_{TC''}$. Let $N^\sim = \overline{\hat{N} - TC''}$, $q^\sim = \hat{q}|_{N^\sim} : N^\sim \longrightarrow C''$, $\pi^\sim = h_S \cdot q^\sim : N^\sim \longrightarrow S''$,

$C^{\sim}$ be the cross-section at infinity of $\hat{q}: \hat{N} \longrightarrow C''$, that is, $C^{\sim} = \hat{N}-N$.

Since V_1 is transversal to $\partial\tilde{T}$, $\tilde{\tau}|_{V_1\cap\tilde{\tau}^{-1}(S'')}: V_1\cap\tilde{\tau}^{-1}(S'') \longrightarrow S''$ is a differential fibre bundle. (Changing if necessary $\tilde{T}$ we can assume that $\tilde{\tau}|_{\tilde{\tau}^{-1}(S'')}: \tilde{\tau}^{-1}(S'') \longrightarrow S''$ is a 4-disk bundle). Using the same arguments as in ([10],Lemma 2.1) we obtain that the typical fiber of $\tilde{\tau}|_{V_1\cap\tilde{\tau}^{-1}(S'')}$ is diffeomorphic to $S^1 \times I$ $(I=\{x\in\mathbb{R}\,|\,0\leq x\leq 1\})$. Using our identification of $\overline{V_1-V_1\cap\tilde{T}}$ with $\tilde{V}_o-h^{-1}(V_o\cap\tilde{T})$ and the embedding $TC'' \subset \hat{N}$ we obtain a diffeomorphism $\psi^{\sim}: \partial(V_1 \cap \tilde{\tau}^{-1}(S'')) \longrightarrow \partial N^{\sim}$. We can assume that $\tilde{\tau}|_{\partial(V_1\cap\tilde{\tau}^{-1}(S''))} = (\pi^{\sim}|_{\partial N^{\sim}})\cdot\psi^{\sim}$. Now let $A^{\sim} = [V_1\cap\tilde{\tau}^{-1}(S'')]\cup_{\psi^{\sim}} N^{\sim}$, and $\pi^{\approx}: A^{\sim} \longrightarrow S''$ be defined by

$$\pi^{\approx}(x) = \begin{cases} \tilde{\tau}(x) \text{ if } x \in V_1 \cap \tilde{\tau}^{-1}(S'') \\ \pi^{\sim}(x) \text{ if } x \in N^{\sim}. \end{cases}$$

Because $\pi^{\sim}: N^{\sim} \longrightarrow S''$ is a fibre bundle with typical fiber diffeomorphic to the disjoint union of two closed 2-disks, we see that $\pi^{\approx}: A^{\sim} \longrightarrow S''$ is an S^2-bundle.

It is easy to verify that we can do all our identifications and constructions so that for any (ℓ,i), $\ell = 1,2,\cdots,\nu$, $i = 1,2,3$,

$$\overline{M}_{\ell,i}|_{K_{\ell,i}-\partial d^{\ell}_{io}} \longrightarrow K_{\ell,i} - \partial d^{\ell}_{io},\ \ell = 1,2,\cdots,\nu,\ i = 1,2,3,$$

will coincide with

$$\widetilde{\pi}\big|_{(\widetilde{\pi})^{-1}(K_{\ell,i}-\partial d^{\ell}_{io})} : (\widetilde{\pi})^{-1}(K_{\ell,i}-\partial d^{\ell}_{io}) \to K_{\ell,i}-\partial d^{\ell}_{io},$$

$C_{\ell,ji} \cap C_{\ell,ki}$ (where $(j,k) = (1,2,3)-(i)$) will coincide with $(\widetilde{\pi})^{-1}(K_{\ell,i}-\partial d^{\ell}_{io}) \cap C^{\sim}$ and for any $m = 1,2,\cdots,\rho$

$\overline{\pi}_m|_{\overline{\pi}_m^{-1}(K_m-d_{mo})} : \overline{\pi}_m^{-1}(K_m-d_{mo}) \to K_m-d_{mo}$ will coincide with

$\widetilde{\pi}\big|_{(\widetilde{\pi})^{-1}(K_m-d_{mo})} : (\widetilde{\pi})^{-1}(K_m-d_{mo}) \longrightarrow K_m-d_{mo}$, and

$C_m \cap \overline{\pi}_m^{-1}(K_m-d_{mo})$ will coincide with $C^{\sim} \cap (\widetilde{\pi})^{-1}(K_m-d_{mo})$.

We define $\hat{A}$ as union of $A^{\sim}, \hat{P}_1,\cdots,\hat{P}_\nu,\overline{T}_1,\cdots,\overline{T}_\rho$ where we identify a point $x \in (\widetilde{\pi})^{-1}(K_{\ell,i}-\partial d^{\ell}_{io})(\subset A^{\sim})$, $\ell = 1,2,\cdots,\nu$, $i = 1,2,3$, with the corresponding point $x' \in \overline{M}_{\ell,i}|K_{\ell,i}-\partial d^{\ell}_{io}$ $(\subset P_\ell)$ and a point $y \in (\widetilde{\pi})^{-1}(K_m-d_{mo})(\subset\hat{A})$ with the corresponding point $y' \in \overline{\pi}_m^{-1}(K_m-d_{mo})(\subset\overline{T}_m)$.

Let $\hat{C} = C^{\sim} \cup [(\bigcup_{\ell=1}^{\nu} C^{(\ell)}) \cup (\bigcup_{m=1}^{\rho} C_m]$ (union in $\hat{A}$),

$\hat{TC} = N^{\sim} \cup [(\overline{\bigcup_{\ell=1}^{\nu}(\hat{P}_\ell-\hat{\lambda}_\ell(\hat{Z}_\ell))}) \cup (\bigcup_{m=1}^{\rho} J_m)]$ (union in $\hat{A}$).

Because each $\overline{M}_{\ell,i} \longrightarrow K_{\ell,i}$, $\ell = 1,2,\cdots,\nu$, $i = 1,2,3$, is a trivial S^2-bundle and $C_{\ell,ji}, C_{\ell,ki}$ are "horizontal" cross-sections in $\overline{M}_{\ell,i} \longrightarrow K_{\ell,i}$, we can extend $\widetilde{\pi}: A^{\sim} \longrightarrow S"$ to an S^2-bundle $\widetilde{\widetilde{\pi}}: A^{\approx} \longrightarrow \widetilde{S} - \bigcup_{m=1}^{\rho} d_{mo}$ and $C^{\sim}$ to a 2-manifold $C^{\approx}$ in $A^{\approx}$ where for any (ℓ,i), $\ell = 1,2,\cdots,\nu$, $i = 1,2,3$, $C^{\approx} \cap (\widetilde{\widetilde{\pi}})^{-1}(\tilde{d}^{\ell}_{io})$ is equal to disjoint union of two cross-sections of $\tilde{\tilde{\pi}}$ over $\tilde{d}^{\ell}_{io}$.

Using the canonical embedding of $\bar{\pi}_m^{-1}(K_m) \longrightarrow K_m$ in $\bar{\pi}_m: \bar{T}_m \longrightarrow d_m$ we extend $\tilde{\tilde{A}}$ to a differential manifold A' $(A'-\tilde{\tilde{A}} = \bigcup_{m=1}^{\rho} \bar{\pi}_m^{-1}(d_{mo}))$, the map $\tilde{\tilde{\pi}}$ to a differential map

$$\bar{\pi}': A' \longrightarrow \tilde{S} \; (\pi'|_{\tilde{\tilde{A}}} = \tilde{\tilde{\pi}}, \quad \pi'|_{\bar{\pi}_m^{-1}(d_{mo})} = \bar{\pi}_m|_{\bar{\pi}_m^{-1}(d_{mo})}$$

and $\tilde{\tilde{C}}$ to a 2-manifold L' in A' $(L' \cap \tilde{\tilde{A}} = \tilde{\tilde{C}}$, $L' \cap \bar{\pi}_m^{-1}(d_{mo}) = C_m \cap \bar{\pi}_m^{-1}(d_{mo}))$. The construction of $\bar{\pi}_m: \bar{T}_m \longrightarrow d_m$ shows that we can write $\pi' = \pi \cdot \varphi$, where $\pi: A \longrightarrow \tilde{S}$ is an S^2-bundle, there exists a differential embedding $i_C: \tilde{C} \longrightarrow A$ such that $\pi \cdot i_C: \tilde{C} \longrightarrow \tilde{S}$ coincides with canonical map corresponding to $h|_C: C \longrightarrow S$, $\varphi: A' \longrightarrow A$ is the monoidal transformation of A with center equal to $\bigcup_{m=1}^{\rho} q_m$, where $q_m = \pi^{-1}(h_S^{-1}(y_m)) \cap i_C(\tilde{C})$ and L' is equal to the strict image of $i_C(\tilde{C})$ in A'.

It is easy to verify that we can construct $\hat{A}$, $\hat{C}$ and $T\hat{C}$ from A' and L' by the same way as in the statements c),d),e) of our theorem. Now the existence of $\eta: \partial TC \longrightarrow \partial T\hat{C}$ with the properties formulated in the statements e) and f) and of diffeomorphisms $u_t: \overline{\tilde{V}_o - TC} \cup_{\eta} \overline{(\hat{A}-T\hat{C})} \longrightarrow V_t$ (see the statement g) of the theorem) easily follows from our constructions. (Recall that above we identified $\overline{V_1 - V_1 \cap \tilde{T}}$ with $\overline{\tilde{V}_o - h^{-1}(V_o \cap \tilde{T})}$ and that $V_1 \cap \tilde{T} = [V_1 \cap \tilde{\tau}^{-1}(S'')] \cup (\bigcup_{\ell=1}^{\nu} \hat{Z}_{\ell}) \cup (\bigcup_{m=1}^{\rho} T_m))$. Q.E.D.

§2. A topological comparison theorem for elements of linear systems on complex threefolds.

Lemma 2. Let $\pi: X \longrightarrow Y_g$ be an oriented differential S^2-bundle, Y_g be a closed oriented surface of genus g, γ_1, γ_2 be two smooth embeddings of S^1 in X such that $\gamma_1 \cap \gamma_2 = \emptyset$, γ_i is a cross-section of π over $\pi(\gamma_i)$ and $\pi(\gamma_1), \pi(\gamma_2)$ intersect transversally at one point of Y_g. Let d be a closed 2-disk in $Y_g - \bigcup_{i=1}^{2} \pi(\gamma_i)$ and $\tilde{X}$ be obtained from X by the surgeries along γ_1 and γ_2. Let Z be the image of $\pi^{-1}(d)$ in $\tilde{X}$. Then there exist a differential map $\tilde{\pi}: \tilde{X} \longrightarrow Y_{g-1}$ and a 2-disk $\tilde{d} \subset Y_{g-1}$ such that

(a) $\tilde{\pi}: \tilde{X} \longrightarrow Y_{g-1}$ is a differentiable S^2-bundle,

(b) $Z = \tilde{\pi}^{-1}(\tilde{d})$ and $\tilde{\pi}|_{\tilde{\pi}^{-1}(\tilde{d})}: \tilde{\pi}^{-1}(\tilde{d}) \longrightarrow \tilde{d}$ coincide with $Z \longrightarrow d$ corresponding to $\pi|_{\pi^{-1}(d)}: \pi^{-1}(d) \longrightarrow d$.

Proof. Using the triviality of the S^2-bundle over S^1 and considering a tubular neighborhood of $\bigcup_{i=1}^{2} \pi(\gamma_i)$ in Y_g we see that only the case which we have to look at is the case $g = 1$. Thus assume $g = 1$. Let p be the center of d, $S_p^2 = \pi^{-1}(p)$ and X' be obtained from X by the surgery along S_p^2. Let γ_1', γ_2' be the images of γ_1, γ_2 in X' and $\tilde{X}'$ be obtained from X' by surgeries along γ_1', γ_2'. Using uniqueness of tubular neighborhood (of S^1 in S^4) we have to prove only that $\tilde{X}'$ is diffeomorphic to S^4.

First of all we shall prove that there exists a diffeomorphism $\alpha\colon X' \longrightarrow (S^1 x S^3)_1 \# (S^1 x S^3)_2$ (connected sum of copies of $S^1 x S^3$) such that for $i = 1,2$ $\alpha(Y_i) = (S^1 x a_i)_i \subset (S^1 x S^3)_i$, $a_1, a_2 \in S^3$.

Let $Y_i^o = \pi(Y_i)$, $i = 1,2$, $q = Y_1^o \cap Y_2^o$ and B be a small closed 2-disk in Y_1-p with center q. There exist differential embeddings:

$$\psi_i\colon I \times I \longrightarrow Y_1\text{-}p - \text{int } B, \quad i = 1,2$$

such that $\psi_1(I x I) \cap \psi_2(I x I) = \emptyset$,

$$\psi_i(I x \partial I) \subset \partial B, \quad \psi_i(\tfrac{1}{2} x I) = \overline{Y_i - Y_i \cap B}, \ i = 1,2 \text{ (see Fig. 4)}.$$

Denote $A_i = \psi_i(I x I)$, $A = A_1 \cup A_2 \cup B$.

We can assume that $\pi^{-1}(A) \subset X'$, $X' - \pi^{-1}(A) = S^1 x D^3$ and the structures of $S^1 x S^2$ on $\pi^{-1}(\partial A)$ obtained correspondingly from $\pi^{-1}(\partial A) \longrightarrow \partial A$ and from $\partial(X' - \pi^{-1}(A)) = \partial(S^1 x D^3)(= S^1 x S^2)$ are the same. Let $\tau\colon X' - \pi^{-1}(A) \longrightarrow \partial A$ be a projection corresponding to the equality $X' - \pi^{-1}(A) = S^1 x D^3$. We have

$$[\pi^{-1}(\psi_i(I x \tfrac{1}{2})) \underset{\text{(union in X')}}{\cup} \tau^{-1}(\psi_i(\partial I + \tfrac{1}{2}))] \ x \ I \approx [(I x S^2) \cup_{S^o x S^2} (S^o x D^3)] x I$$

$$\approx S^3 \ x \ I.$$

It follows from this that $\pi^{-1}(\psi_i(I x I)) \cup \tau^{-1}(\psi_i(\partial I x I))$ (union in X') is diffeomorphic to $S^3 x I$.

Fig. 4

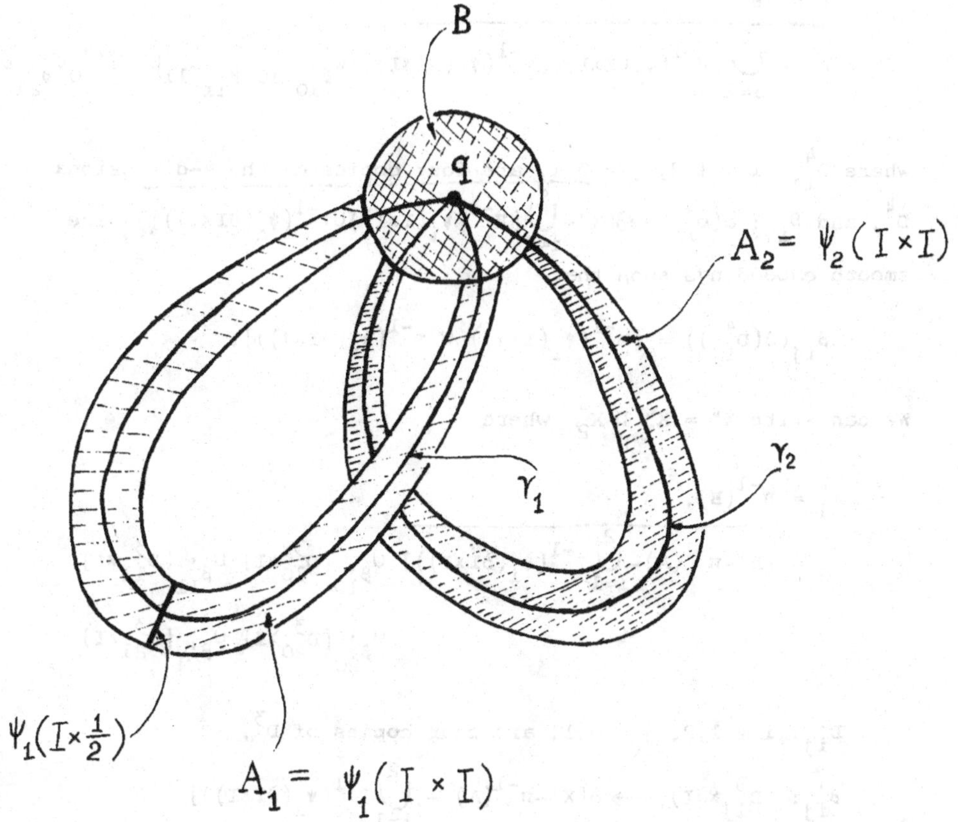

Let X" be obtained from X' by the surgeries along $\pi^{-1}(\psi_i(I\times\frac{1}{2})) \cup \tau^{-1}(\psi_i(\partial I\times\frac{1}{2}))(\approx S^3)$, i = 1,2.

We have

$$X" = \overline{X' - \bigcup_{i=1}^{2}[\pi^{-1}(\psi_i(I\times I))\cup\tau^{-1}(\psi_i(\partial I\times I))]} \cup_{\beta_{10}} D^4_{10} \cup_{\beta_{11}} D^4_{11} \cup_{\beta_{20}} D^4_{20} \cup_{\beta_{21}} D^4_{21}$$

where D^4_{ij}, i = 1,2, j = 0,1, are four copies of the 4-dimensional ball D^4, and $\beta_{ij}\colon \partial(D^4_{ij}) \to \partial(\overline{X' - \bigcup_{i=1}^{2}[\pi^{-1}(\psi_i(I\times I))\cup\tau^{-1}(\psi_i(\partial I\times I))]})$ are smooth embeddings such that

$$\beta_{ij}(\partial(D^4_{ij})) = [\pi^{-1}(\psi_i(I\times j)) \cup \tau^{-1}(\psi_i(\partial I\times j))].$$

We can write $X" = X"_1 \cup_{\eta} X"_2$ where

$$X"_1 = \pi^{-1}(B),$$

$$X"_2 = \overline{[X'-\pi^{-1}(A)-\bigcup_{i=1}^{2}\tau^{-1}(\psi_i(\partial I\times I))]} \cup_{\beta'_{10}} [D^3_{10}\times I] \cup_{\beta'_{11}} [D^3_{11}\times I]$$

$$\cup_{\beta'_{20}} [D^3_{20}\times I] \cup_{\beta'_{21}} [D^3_{21}\times I],$$

D^3_{ij}, i = 1,2, j = 0,1, are four copies of D^3,

$$\beta'_{ij}\colon (D^3_{ij}\times\partial I) \longrightarrow \partial[X'-\pi^{-1}(A) - \bigcup_{i=1}^{2}\tau^{-1}(\psi_i(\partial I\times I))]$$

are smooth embeddings such that

$$\beta'_{ij}(D^3_{ij}\times k) = \tau^{-1}\psi_i(k,j), \qquad i = 1,2, \quad j = 0,1, \quad k = 0,1,$$

and $\eta\colon \partial X"_1 \longrightarrow \partial X"_2$ is a diffeomorphism such that

$$\eta\,\pi^{-1}\overline{\left(\partial B-\bigcup_{i=1}^{2}\psi_i(I\times\partial I)\right)} = \partial\overline{\left(X'-\pi^{-1}(A) - \bigcup_{i=1}^{2}\tau^{-1}(\psi_i(\partial I\times I)\right)},$$

$$\eta\big|_{\pi^{-1}\overline{(\partial B-\bigcup_{i=1}^{2}\psi_i(I\times\partial I)}}: \pi^{-1}\overline{(\partial B-\bigcup_{i=1}^{2}\psi_i(I\times\partial I)} \longrightarrow$$

$$\longrightarrow \partial\overline{(X' - \pi^{-1}(A) - \bigcup_{i=1}^{2}\tau^{-1}(\psi_i(\partial I\times I))}$$

is the identity map,

$$\eta(\pi^{-1}(\psi_i(I\times j)) = \partial D^3_{ij}\times I, \quad i = 1,2, \quad j = 0,1,$$

and $\forall\ t \in I$

$$\eta(\pi^{-1}(\psi_i(t\times j)) = \partial D^3_{ij} \times t.$$

We see that we can consider X_1'' as $D^2\times S^2$, X_2'' as $S^1\times D^3$ and η as natural identification of $\partial(D^2\times S^2)$ and $\partial(S^1\times D^3)$. Thus X" is diffeomorphic to S^4.

Our construction of X" shows that we can get X' from S^4 performing surgeries along two embeddings of S^o in S^4 (see Fig. 5), say (b_{10}, b_{11}) and (b_{20}, b_{21}), such that Y_1' and Y_2' are obtained as follows (Fig. 5).

Let D^4_{ij}, $i = 1,2$, $j = 0,1$, be small balls with the centers b_{ij}, c_{ij} be points in $\partial(D^4_{ij})$, δ_i be smooth disjoint paths connecting c_{io} with c_{i1}, in $S^4 - \bigcup_{i=1}^{2}\bigcup_{j=0}^{1} D^4_{ij}$, $\partial D^4_{ij} = S^3_i \times j$, ϵ_i be smooth paths in $S^3_i\times I$ which are cross-sections of $S^3_i\times I \longrightarrow I$ and connect c_{io} with c_{i1} in $S^3_i\times I$. Then $Y_1' = \delta_1\cup\epsilon_1$, $Y_2' = \delta_2\cup\epsilon_2$.

Fig. 5

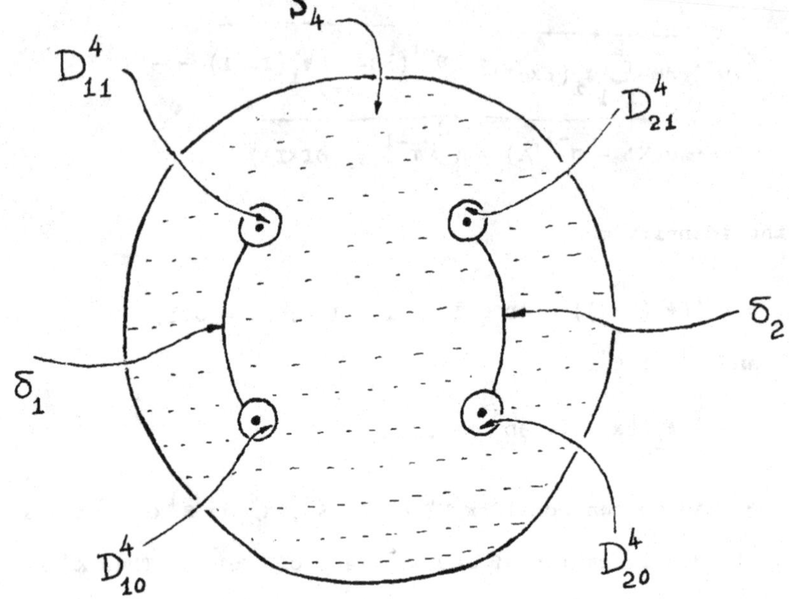

We see from this that there exists a diffeomorphism

$\alpha\colon X' \longrightarrow (S^1 x S^3)_1 \# (S^1 x S^3)_2$ such that

$\alpha(Y_i') = (S^1 x a_i)_i \subset (S^1 x S^3)_i$, $i = 1,2$, $a_1, a_2 \in S^3$.

From $S^1 x S^3 = (S^1 x S^3_+) \cup_{S^1 x S^2} (S^1 x S^3_-)$ it follows that a surgery on $S^1 x S^3$ along $S^1 x a$, $a \in S^3$, transforms $S^1 x S^3$ in S^4. This immediately gives that surgery on X' along Y_1' and Y_2' transforms X' into a connected sum of two copies of S^4, that is, into S^4. Q.E.D.

<u>Theorem 2</u>. Let W be a 3-dimensional compact complex manifold, [E] be a complex (analytic) line bundle on W, φ_0 and φ_1 be two global (holomorphic) cross-sections of [E]. Suppose that the zero-divisor $(\varphi_1)_0$ of φ_1 is a complex submanifold V_1 of W, the singular locus of the zero-divisor $(\varphi_0)_0 = V_0$ of φ_0 is canonical and the locus S of all ordinary singularities of V_0 is an irreducible complex curve. Suppose also that V_1 is transversal to V_0 in the following sense: $\forall$ $x \in V_1 \cap V_0$ there exists a local complex coordinate system (z_1,z_2,z_3) on W with the center x such that in some neighborhood U_x of x in W, V_1 is defined by the equation $z_3 = 0$ and V_0 is defined either by the equation $z_1 = 0$ or by the equation $z_1 z_2 = 0$. Let $h\colon \overline{V} \longrightarrow V_0$ be a minimal desingularization of V_0. Denote by ρ the number of pinch-points of V_0, by ν the number of triplanar points of V_0, by b the intersection number $S.V_1$ (in W) and by g(S) the genus of S (that is, the genus of a non-singular model of S).

Suppose that $\rho \neq 0$ and that $\pi_1(V_1) = \pi_1(\bar{V}) = 0$. Then

1) if $b \neq 0$, then $V_1 \# P$ is diffeomorphic to
$\bar{V} \# (2\nu+\rho+2g(S)-1)P \# (\nu+2\rho+b+2g(S)-2)Q$ and

2) if $b = 0$ then $V_1 \# P \# Q$ is diffeomorphic to
$\bar{V} \# (2\nu+\rho+2g(S)-1)P \# (\nu+2\rho+2g(S)-1)Q$.

<u>Proof</u>. Let $B = V_o \cap V_1$. We construct a modification $\mathfrak{F}: \tilde{W} \longrightarrow W$ (following Hironaka's idea) as follows. If $B = \emptyset$ then $\tilde{W} = W$ and $\mathfrak{F}$ is the identity map. If $B \neq \emptyset$ and $B \cap S = \emptyset$ then $\mathfrak{F}$ is the monoidal transformation of W with the center B. Now let $B \cap S \neq \emptyset$ and $B \cap S = \{x_1, x_2, \cdots, x_b\}$. We can take a coordinate neighborhood U_k of W with the center in x_k, $k = 1,2,\cdots,b$, with coordinates $z_1^{(k)}, z_2^{(k)}, z_3^{(k)}$ such that $V_1 \cap U_k$ is given by $z_3^{(k)} = 0$ and $V_o \cap U_k$ is given by $z_1^{(k)} \cdot z_2^{(k)} = 0$ (see Fig. 6). Let $B_i^{(k)}$, $i = 1,2$, be complex curves in U_k given by the equations $z_i^{(k)} = 0$, $z_3^{(k)} = 0$, $\mathfrak{F}_{k1}: U_{k1} \longrightarrow U_k$ be the monoidal transformation of U_k with center $B_1^{(k)}$, $\mathfrak{F}_{k2}: U_{k2} \longrightarrow U_{k1}$ be the monoidal transformations of U_{k1} with center $\bar{B}_2^{(k)}$ where $\bar{B}_2^{(k)}$ is the strict image of $B_2^{(k)}$ in U_{k1} and $\mathfrak{F}^{\cdot}: \tilde{W}^{\cdot} \longrightarrow W^{\cdot}$ be the monoidal transformations of $W^{\cdot} = W - \bigcup_{k=1}^{b} x_k$ with the center $B - \bigcup_{k=1}^{b} x_k$. Evidently $\mathfrak{F}_{1k}\mathfrak{F}_{2k}|_{(\mathfrak{F}_{1k}\mathfrak{F}_{2k})^{-1}(U_k - x_k)}$ coincide with $\mathfrak{F}^{\cdot}|_{\mathfrak{F}^{\cdot -1}(U_k - x_k)}$. Using this identification we add to $\mathfrak{F}^{\cdot}: \tilde{W}^{\cdot} \longrightarrow W^{\cdot}$ the disjoint sum $\bigcup(\mathfrak{F}_{1k}\mathfrak{F}_{2k}: U_{2k} \longrightarrow U_k)$ and

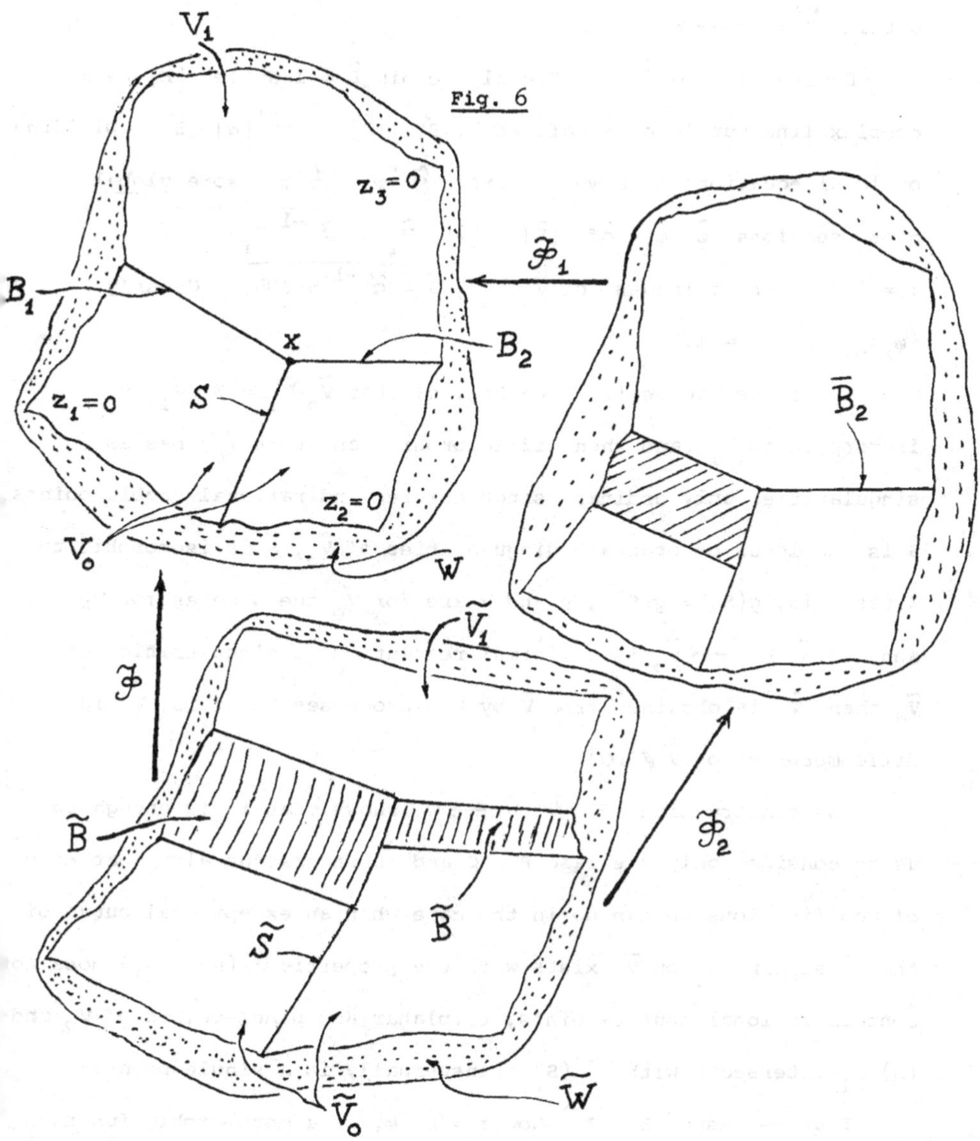
V_1
$z_3=0$
B_1
x
B_2
$z_1=0$
S
$z_2=0$
V_0
W
$\mathfrak{F}_1$
$\bar{B}_2$
$\mathfrak{F}$
$\tilde{V}_1$
$\tilde{B}$
$\tilde{B}$
$\tilde{S}$
$\tilde{V}_0$
$\tilde{W}$
$\mathfrak{F}_2$

Fig. 6

obtain $\mathcal{F}: \tilde{W} \longrightarrow W$.

Denote $\tilde{B} = \mathcal{F}^{-1}(B)$. Clearly $\operatorname{codim}_{\mathbb{C}} \tilde{B} = 1$ so let $[\tilde{B}]$ be a complex line bundle on $\tilde{W}$ defined by $\tilde{B}$, $[\tilde{E}] = \mathcal{F}^*[E]-[\tilde{B}]$. Dividing on local equations of B we get from $\mathcal{F}^*\varphi_o$, $\mathcal{F}^*\varphi_1$ some global cross-sections $\tilde{\varphi}_o, \tilde{\varphi}_1$ of $[\tilde{E}]$. Let $\tilde{V}_i = \overline{\mathcal{F}^{-1}(V_i - V_i \cap B)}$, $i = 1,2$ ("strict images" of V_o, V_1), $\tilde{S} = \overline{\mathcal{F}^{-1}(S - S \cap B)}$. Clearly $(\tilde{\varphi}_i)_o = \tilde{V}_i$, $i = 1,2$.

It is easy to verify (see Fig. 6) that $\tilde{V}_o \cap \tilde{V}_1 = \emptyset$, $\tilde{V}_1$ is isomorphic to V_1 (and then diffeomorphic) and that $\tilde{V}_o$ has as singularities only ordinary singularities and rational double points, $\tilde{S}$ is the locus of ordinary singularities of $\tilde{V}_o$, $\tilde{S}$ is isomorphic to S (that is, $g(\tilde{S}) = g(S)$), ρ and ν are for $\tilde{V}_o$ the same as for V_o and if $\tilde{h}: \bar{\bar{V}} \longrightarrow \tilde{V}_o$ is a minimal resolution of singularities of $\tilde{V}_o$ then $\bar{\bar{V}}$ is obtained from $\bar{V}$ by b σ-processes (that is, $\bar{\bar{V}}$ is diffeomorphic to $\bar{V} \# bQ$).

The construction of $\mathcal{F}: \tilde{W} \longrightarrow W$ shows that it is enough for us to consider only the case $B = \emptyset$ and to understand also what kind of modifications we can do in the case when an exceptional curve of the first kind S_1 on $\bar{V}$ exists with the properties: (a) $h(S_1)$ does not contain rational double-points, triplanar and pinch-points of V_o and (b) S_1 intersects with $h^{-1}(S)$ transversally in a single point.

Thus we assume $B = \emptyset$. Now $f = \varphi_o/\varphi_1$ is a meromorphic function with no indeterminacy points and we can apply Theorem 1. We shall use the same notations as in the formulation of Theorem 1. Using part f) of Theorem 1 we can find a non-singular point $p \in U$, small closed 2-disk d_p in U such that if $\hat{d}_p$ denotes $\eta_o(d_p)$, then we have a commutative diagram:

$$\begin{array}{ccc} \underline{\tau}^{-1}(d_p) & \xrightarrow{\ \eta|_{\underline{\tau}^{-1}(d_p)}\ } & \hat{\underline{\tau}}^{-1}(\hat{d}_p) \\ \Big\downarrow \underline{\tau} & & \Big\downarrow \hat{\underline{\tau}} \\ d_p & \xrightarrow{\ \eta_o|_{d_p}\ } & \hat{d}_p \end{array}$$

Identify now V_1 with $\overline{\bar{V}-TC} \cup_{\eta} (\widehat{\hat{A}-T\hat{C}})$ (that is, we shall consider $\overline{\bar{V}-TC}$ and $\widehat{\hat{A}-T\hat{C}}$ as subspaces of V_1 and η as the identity map). We can construct two embeddings $\psi: (d_p \times [-1,0]) \times S^1 \longrightarrow \overline{\bar{V}-TC}$ and $\hat{\psi}: (\hat{d}_p \times [0,1]) \times S^1 \longrightarrow \widehat{\hat{A}-T\hat{C}}$ such that

$\psi((d_p \times [-1,0]) \times S^1) \cap \partial TC = \psi(d_p \times 0 \times S^1) = \underline{\tau}^{-1}(d_p) = \hat{\underline{\tau}}^{-1}(\hat{d}_p) =$
$\hat{\psi}(\hat{d}_p \times 0 \times S^1) = \hat{\psi}((\hat{d}_p \times [0,1]) \times S^1) \cap \partial T\hat{C}$ and

$\forall\ x \in d_p,\ x' \in \hat{d}_p,\ y \in S^1 \qquad \tau\psi(x \times 0 \times y) = x,\ \hat{\tau}\hat{\psi}(x' \times 0 \times y) = x',$
$\psi(x \times 0 \times y) = \hat{\psi}(\eta_o(x) \times 0 \times y).$

Let $c_p = \partial d_p$, $\hat{c}_p = \partial \hat{d}_p$, $S^2_{p-} = (d_p \times (-1)) \cup (c_p \times [-1,0])$ (union in $d_p \times [-1,0]$),

$S^2_{p+} = (\hat{c}_p \times [0,1]) \cup (\hat{d}_p \times (1))$ (union in $\hat{d}_p \times [0,1]$). We have $c_p \times 0 = \partial S^2_{p-}$, $\hat{c}_p \times 0 = \partial S^2_{p+}$ and $((\eta_o|_{c_p}) \times id)(c_p \times 0) = \hat{c}_p \times 0$.

Define $S^2_p = S^2_{p-} \cup_{(\eta_o|_{c_p} \times id)} S^2_{p+}$ and let $Y'_p = \psi(d_p \times [-1,0] \times S^1)$,

$Y_p'' = \hat{\psi}(\hat{d}_p \times [0,1] \times S^1)$, $Y_p = Y_p' \cup Y_p''$. Clearly we can consider Y_p as a tubular neighborhood in V_1 of $\psi(p \times 0 \times S^1) = \underline{\tau}^{-1}(p)$. Let $\tilde{V}$ be a 4-manifold obtained from V_1 by surgery along $\tau^{-1}(p)$.

We have a map $e: S_p^2 \times S^1 \longrightarrow \partial Y_p$ defined as follows:

$$e(x,y) = \begin{cases} \hat{\psi}(x,y) & \text{if } x \in S_{p+}^2 \text{ and} \\ \psi(x,y) & \text{if } x \in S_{p-}^2. \end{cases}$$

Now $\tilde{V} = (S_p^2 \times D^2) \cup_e (\overline{V_1 - Y_p})$. We can decompose $\tilde{V}$ as follows: $\tilde{V} = X_1 \cup X_2$ where

$$X_1 = (S_{p-}^2 \times D^2) \cup_{e|S_{p-}^2 \times S^1} (\overline{\overline{V}_0 - TC - Y_p'}),$$

$$X_2 = (S_{p+}^2 \times D^2) \cup_{e|S_{p+}^2 \times S^1} (\overline{\hat{A} - T\hat{C} - Y_p''}).$$

We can construct "diffeomorphisms" $\varphi': \overline{\overline{V} - TC - Y_p'} \longrightarrow \overline{\overline{V} - TC}$, $\varphi'': \overline{\hat{A} - T\hat{C} - Y_p''} \longrightarrow \overline{\hat{A} - T\hat{C}}$ such that φ' (corresp. φ'') is the identity outside of some small neighborhood of

$\psi(S_{p-}^2 \times S^1)$ (corresp. $\hat{\psi}(S_{p+}^2 \times S^1)$, $\varphi'(\psi(S_{p-}^2 \times S^1)) = \psi(d_p \times 0 \times S^1)$ (corresp. $\varphi''(\hat{\psi}(S_{p+}^2 \times S^1) = \hat{\psi}(\hat{d}_p \times 0 \times S^1))$ and there exist "diffeomorphisms" $\underline{\varphi}': S_{p-}^2 \longrightarrow d_p$, $\underline{\varphi}'': S_{p+}^2 \longrightarrow \hat{d}_p$ such that $\forall\, x \in S_{p-}^2$, $x' \in S_{p+}^2$, $y \in S^1$ we have

$\varphi'(\psi(x,y)) = \psi(\underline{\varphi}'(x) \times 0 \times y)$, $\varphi''(\hat{\psi}(x',y) = \hat{\psi}(\underline{\varphi}''(x') \times 0 \times y)$.

Now let $C^\cdot = \overline{C - d_p}$, $\hat{C}^\cdot = \overline{\hat{C} - \hat{d}_p}$, $TC^\cdot = \tau^{-1}(C^\cdot)$, $T\hat{C}^\cdot = \hat{\tau}^{-1}(\hat{C}^\cdot)$.

We can write $\overline{V} - TC^\cdot = (d_p \times D^2) \cup_{\alpha'} (\overline{V} - TC)$, $\overline{\hat{A} - T\hat{C}^\cdot} = (\hat{d}_p \times D^2) \cup_{\alpha''} \overline{(\hat{A} - T\hat{C})}$

where

$$\alpha': d_p \times S^1 \longrightarrow \underline{\tau}^{-1}(d_p), \quad \alpha'': \hat{d}_p \times S^1 \longrightarrow \hat{\underline{\tau}}^{-1}(\hat{d}_p)$$

are some trivializations of $\underline{\tau}^{-1}(d_p) \longrightarrow d_p$ and $\hat{\underline{\tau}}^{-1}(\hat{d}_p) \longrightarrow \hat{d}_p$.

Let $e': S^2_{p-} \times S^1 \longrightarrow \Psi(d_p \times 0 \times S^1)$ (corresp. $e'': S^2_{p+} \times S^1 \to \hat{\Psi}(\hat{d}_p \times 0 \times S^1)$) be equal to

$$(\varphi'|_{\Psi(S^2_{p-} \times S^1)}) \circ (e|_{S^2_{p-} \times S^1}) \text{ (corresp. } (\varphi''|_{\hat{\Psi}(S^2_{p+} \times S^1)}) \cdot (e|_{S^2_{p+} \times S^1})).$$

We have diffeomorphisms

$$X_1 \longrightarrow (S^2_{p-} \times D^2) \cup_{e'} (\overline{\bar{V} - TC}),$$

$$X_2 \longrightarrow (S^2_{p+} \times D^2) \cup_{e''} (\overline{\hat{A} - T\hat{C}}),$$

and because of the commutativity of the diagrams,

$$\begin{array}{ccc} S^2_{p-} \times S^1 & \xrightarrow{e'} & \Psi(d_p \times 0 \times S^1) \\ \downarrow \text{pr} & & \downarrow \underline{\tau} \\ S^2_{p-} & \xrightarrow{\underline{\varphi}'} & d_p \end{array} \quad \text{and} \quad \begin{array}{ccc} S^2_{p+} \times S^1 & \xrightarrow{e''} & \hat{\Psi}(\hat{d}_p \times 0 \times S^1) \\ \downarrow \text{pr} & & \downarrow \hat{\underline{\tau}} \\ S^2_{p+} & \xrightarrow{\underline{\varphi}''} & \hat{d}_p \end{array}$$

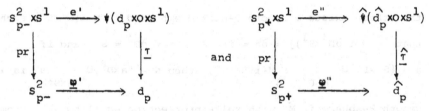

we can identify X_1 with $\overline{\bar{V} - TC^\cdot}$ and X_2 with $\overline{\hat{A} - T\hat{C}^\cdot}$. We see that there exists a diffeomorphism $\eta^\cdot : \partial TC^\cdot \longrightarrow \partial T\hat{C}^\cdot$ such that $\tilde{V}$ is diffeomorphic to $\overline{\bar{V} - TC^\cdot} \cup_{\eta^\cdot} \overline{\hat{A} - T\hat{C}^\cdot}$. We see also that outside of some small neighborhoods of $\underline{\tau}^{-1}(d_p)$ and $\hat{\underline{\tau}}^{-1}(\hat{d}_p)$ $\eta^\cdot$ coincides with η. In particular, we have that $\eta^\cdot$ reverses orientations coming from orientations on $TC^{\cdot}$ and $\widehat{TC}^{\cdot}$ defined by complex structures of $\overline{V}$ and A'.

Suppose that an exceptional curve of the first kind S_1 with the properties mentioned above exists on $\overline{V}$. Let $\sigma: \overline{V} \longrightarrow \underline{V}$ be the corresponding contraction. Without loss of generality we can assume that $p = h^{-1}(s) \cap S_1$ $(= C \cap S_1)$ and $S^1 \cap TC = \tau^{-1}(p)$.

Let $\underline{C} = \sigma(C)$, $\underline{C}^{\cdot} = \sigma(C^{\cdot})$ and $T\underline{C}^{\cdot} = \sigma(TC^{\cdot})$. We see that $S_1 \cap TC^{\cdot} = \emptyset$, that is, we can identify $TC^{\cdot}$ with $T\underline{C}^{\cdot}$ and construct S_1 $(\subset \overline{V} - TC^{\cdot}(\subset \widetilde{V}))$ in $\widetilde{V}$. Let $\tilde{\sigma}: \widetilde{V} \longrightarrow \underline{\widetilde{V}}$ be the corresponding contraction. We see that $\underline{\widetilde{V}}$ is diffeomorphic to $\overline{(\underline{V} - T\underline{C}^{\cdot})} \cup_{\eta^{\cdot}} \overline{(\hat{A} - T\hat{C}^{\cdot})}$.

Now let us make the following remark. Let M be a simply-connected 4-manifold, $i: D^3 \times S^1 \longrightarrow M$ be a smooth embedding. Suppose that there exists a smooth 2-disk d embedded in $M - i(D^3 \times S^1)$ such that $d \cap i(\partial D^3 \times S^1)) = \partial d = i(a \times S^1)$, $a \in \partial D^3 = S^2$ and if $\widetilde{M} = (S^2 \times D^2) \cup_{i|\partial(D^3 \times S^1)} \overline{M - i(D^3 \times S^1)}$ then $N = (a \times D^2) \cup_{i|a \times S^1} d$ is a smooth 2-sphere in $\widetilde{M}$ with self-intersection equal to -1. From the last condition it easily follows that $\widetilde{M}$ is diffeomorphic to $M \# P \# Q$ and if $\sigma_N: \widetilde{M} \longrightarrow \underline{M}$ is the contraction of N to a point then $\underline{M}$ is diffeomorphic to $M \# P$.

Applying this remark to the case $M = V_1$, $N = S_1$ we see that $\underline{\widetilde{V}}$ is diffeomorphic to $V_1 \# P$. We use now the notations $\overline{\overline{V}}$ and $\overline{C}$ for $\underline{V}$ and $\underline{C}$ in the case when there exists S_1 with the properties formulated above and for $\overline{V}$ and C in the case when such S_1 does not exist. Our aim now will be the prove the following

Statement (*): $(\overline{\overline{V}-T\overline{C}^{\cdot}})\cup_{\eta^{\circ}}(\widehat{\hat{A}-T\hat{C}^{\cdot}})$ is diffeomorphic to

$\overline{\overline{V}} \# (2\nu+\rho+2g(S)-1)P \# (\nu+2\rho+2g(S)-1)Q.$

Before proving the Statement (*) let us show how our Theorem follows from it.

Case 1). $b \neq 0$. We have $\overline{\overline{V}} = \underline{V}$ and Statement (*) says that $\underset{\sim}{\tilde{V}}$ is diffeomorphic to $\underline{V} \# (2\nu+\rho+2g(S)-1)P \# (\nu+2\rho+2g(S)-1)Q$. Going back to our arguments with $\mathfrak{F} : \tilde{W} \longrightarrow W$ (in the beginning of proof of the theorem) we see that $\underline{V}$ is diffeomorphic in our initial notations to $\overline{V} \# (b-1)Q$. Because $\tilde{V}_1$ is diffeomorphic to $V_1 \# P$ we get a diffeomorphism

$$V_1 \# P \approx \overline{V} \# (2\nu+\rho+2g(S)-2)P \# (\nu+2\rho+b+2g(S)-2)Q.$$

Case 2). $b = 0$. Let $S^2 \underset{\sim}{\times} S^2$ be $S^2 \times S^2$ or $P \# Q$. We have $\overline{\overline{V}} = \overline{V}$, $\tilde{V} \approx V_1 \# S^2 \underset{\sim}{\times} S^2$ and Statement (*) says that

$$V_1 \# S^2 \underset{\sim}{\times} S^2 \approx \overline{V} \# (2\nu+\rho+2g(S)-1)P \# (\nu+2\rho+2g(S)-1)Q.$$

Note that ρ is the number of branch points for the map $\tilde{C} \longrightarrow \tilde{S}$ where $\tilde{C},\tilde{S}$ are normalizations for C and S and $\tilde{C} \longrightarrow \tilde{S}$ is induced by $h|_C: C \longrightarrow S$. Hence ρ is even and the condition $\rho \neq 0$ really means $\rho \geq 2$. We see that $\nu+2\rho+2g(S)-2 > 0$. Hence, $V_1 \# S^2 \underset{\sim}{\times} S^2$ has odd intersection form. Suppose $S^2 \underset{\sim}{\times} S^2$ is actually $S^2 \times S^2$.

Then if V_1 has odd intersection form we have by a result of Wall (see [8]) that $V_1 \# S^2 x S^2 \approx V_1 \# P \# Q$ and if V_1 has even intersection form then $V_1 \# S^2 x S^2$ has also even intersection form which contradicts our remark above. We see that always we can write here $V_1 \# S^2 \underset{\sim}{x} S^2 \approx V_1 \# P \# Q$ and we get

$$V_1 \# P \# Q \approx \bar{V} \# (2\nu+\rho+2g(S)-1)P \# (\nu+2\rho+2g(S)-1)Q.$$

Proof of Statement (*).

Let $S' = \tilde{S} - \bigcup_{\ell=1}^{\nu} \bigcup_{i=1}^{3} d_i^{\ell}$, $C' = \pi'^{-1}(S') \cap L'$, $\pi'' = \pi'|_{C'}: C' \longrightarrow S'$, $g = g(S)$. Suppose $g > 0$. We choose smooth circles α_i, $i = 1,2,\cdots,2g$ on $S' - \bigcup_{m=1}^{\rho} q'_m$, where $q'_1,\cdots,q'_\rho$ are all the branch points of π'' on S', with the following properties: (1) if $|j-i| \neq g,0$ then $\alpha_i \cap \alpha_j = \emptyset$ and (2) if $1 \leq i \leq g$ then α_i intersects α_{i+g} transversally and in a single point. We claim that we always can make our choice such that the following additional property holds: (3) There exist smooth circles α'_i, $i = 1,\cdots,2g$, on C' such that α'_i is a cross-section of $\pi''|_{\pi''^{-1}(\alpha_i)}: \pi''^{-1}(\alpha_i) \longrightarrow \alpha_i$, and for $1 \leq i \leq g$ α'_i intersects α'_{i+g} transversally and in one point. We proceed by induction. Let $\alpha_o = \emptyset$, $\alpha'_o = \emptyset$ and suppose that $\alpha_1,\cdots,\alpha_{2g}$ with the properties (1), (2) are chosen such that when $k \neq 0$ there exist smooth circles α'_i, $i = 1,2,\cdots,k$, on C' such that α'_i is a cross-section of $\pi''|_{\pi''^{-1}(\alpha_i)}: \pi''^{-1}(\alpha_i) \longrightarrow \alpha_i$

and for any pair (i,j), $i,j = 1,2,\cdots,k$ with $|j-i| = g$, α_i' intersects α_j' transversally and in one point (for $k = 0$ these conditions are empty). Consider α_{k+1}. Suppose that $\pi''^{-1}(\alpha_{k+1})$ is connected. Recall that $\rho \neq 0$ and let $q' = q_1'$. Take $a \in \alpha_{k+1}$, $a \notin \bigcup_{\substack{i=1 \\ i\neq k+1}}^{2g} \alpha_i$, and $a' \in \pi''^{-1}(a)$. Let $i'\colon I \longrightarrow C'$ be a path in C' which is the lifting of α_{k+1} to C' satisfying the condition $i'(0) = a'$. Because $\pi''^{-1}(\alpha_{k+1})$ is connected we have $i(1) \neq a'$.

Note that $\alpha_1,\cdots,\alpha_{2g}$ is the so-called canonical basis of S' and we have that $S' - \bigcup_{\substack{i=1 \\ i\neq k+1}}^{2g} \alpha_i$ is connected. Hence we can find a smooth path $\gamma \subset S' - \bigcup_{\substack{i=1 \\ i\neq k+1}}^{2g} \alpha_i$ connecting a with q' and such that $\gamma \cap \alpha_{k+1} = a$. (See Fig. 7). Take a point $b \in \alpha_{k+1}$, $b \neq a$, $b \notin \bigcup_{\substack{i=1 \\ i\neq k+1}}^{2g} \alpha_i$, and let $\delta(a,b)$ be the arc on α_{k+1} connecting a and b such that $\delta_{(a,b)} \cap (\bigcup_{\substack{i=1 \\ i\neq k+1}}^{2g} \alpha_i) = \emptyset$.

We can find a smooth path $\gamma(a,b)$ on $S' - \bigcup_{\substack{i=1 \\ i\neq k+1}}^{2g} \alpha_i - \bigcup_{m=1}^{\rho} q_m'$ connecting a and b and such that $\gamma_{a,b} \cap \alpha_{k+1} = \{a,b\}$, the closed path $\gamma_{(a,b)} \cup \delta_{(a,b)}$ is homotopically trivial on S' and if $\mathcal{V}$ is the domain on S' with $\partial\mathcal{V} = \gamma_{(a,b)} \cup \delta(a,b)$ then $\mathcal{V} \cap (\bigcup_{m=1}^{\rho} q_m') = q'$.

Fig. 7

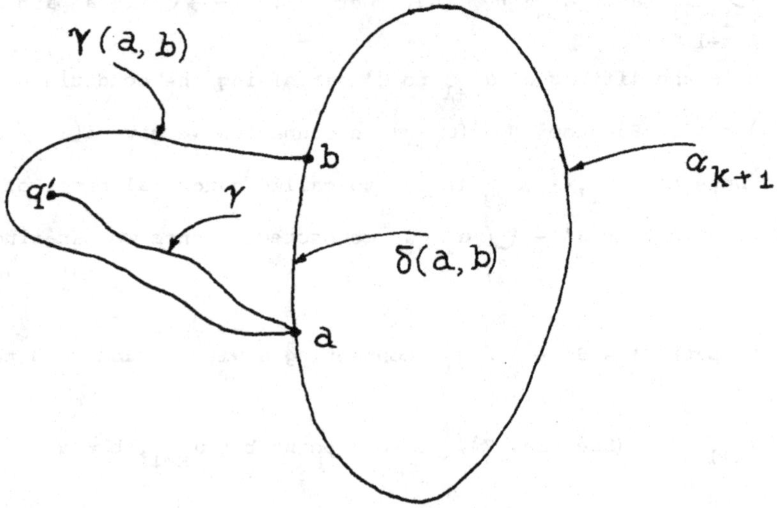

(We choose b and $\gamma(a,b)$ sufficiently close to a and γ). Denote $\bar{\alpha}_{k+1} = \overline{\alpha_{k+1} - \delta(a,b)} \cup \gamma(a,b)$. Because π'' is of degree two we get that if $\bar{i}': I \longrightarrow C'$ is the lifting of $\bar{\alpha}_{k+1}$ to C' satisfying to condition $\bar{i}'(0) = a'$ then $\bar{i}'(1) = a'$, that is, $\pi''^{-1}(\bar{\alpha}_{k+1})$ is not connected. It is clear that $\alpha_1,\cdots,\alpha_k,\bar{\alpha}_{k+1},\alpha_{k+2},\cdots,\alpha_{2g}$ is again a canonical basis of S'. Because we did not change $\alpha_1,\cdots,\alpha_K$ we see that we could from the beginning assume that $\pi''^{-1}(\alpha_{k+1})$ is not connected. Thus $\pi''^{-1}(\alpha_{k+1})$ has two connected components, say $\alpha_{k+1,1}$ and $\alpha_{k+1,2}$ and if $k+1 > g$ we consider the point

$$x_{k+1} = \pi''^{-1}(\alpha_{k+1} \cap \alpha_{k+1-g}) \cap \alpha'_{k+1-g}$$

and take for α'_{k+1} the one of $\alpha_{k+1,1}$, $\alpha_{k+1,2}$ which contains x_{k+1}. This finishes our induction. Now fix some $\alpha_1,\cdots,\alpha_{2g}$, $\alpha'_1,\cdots,\alpha'_{2g}$ with the properties (1),(2),(3). We can assume also that $\forall\, i = 1,2,\cdots,2g$, $\alpha'_i \subset \hat{C}'$ (that is, $\alpha'_i \cap \hat{d}_p = \emptyset$). If $g = 0$ we take $\alpha_0 = \alpha'_0 = \emptyset$.

Consider P_ℓ, $\ell = 1,2,\cdots,\nu$. Let $\bar{P}_\ell = \overline{P_\ell - \bigcup_{i=1}^{3} T_i^\ell}$. We know that $\hat{C} \cap \bar{P}_\ell = \bigcup_{i=1}^{3} \overline{(E_i^\ell - E_i^\ell \cap T_i^\ell)}$. We can assume that

$E_i^\ell \cap T_i^\ell = \{x \in E_i^\ell \mid \frac{1}{2}\xi_{i''}(x) \le \xi_{i'}(x) \le 2\xi_{i''}(x),\ (i',i'') = (1,2,3)-(i)\}$.

Let $\gamma_{i,i'}^\ell = \{x \in E_i^\ell \mid \xi_{i'}(x) \le \frac{1}{2}\xi_{i''}(x),\ \operatorname{Im}\frac{\xi_{i'}(x)}{\xi_{i''}(x)} = 0,\ i''=(1,2,3)-(i',i)$, $a_{ii'}^\ell$ be a point of P_ℓ with $\xi_i(a_{ii'}^\ell) = 0$, $\xi_{i'}(a_{ii'}^\ell) = \frac{1}{2}$, $\xi_{i''}(a_{ii'}^\ell) = 1$, $\tilde{a}_{ii'}^\ell = (\psi_i^\ell)^{-1}(a_{ii'}^\ell)$.

It is clear that all $\tilde{a}^{\ell}_{ii'}$ are on different connected components of $\partial C'$. Now $\alpha'_1,\cdots,\alpha'_{2g}$ is a part of some canonical basis of C', say $\alpha'_1,\cdots,\alpha'_{2g},\alpha'_{2g+1},\cdots,\alpha'_{2g_C}$, where $g_C = g(C')(= g(\tilde{C})$ where $\tilde{C}$ is the non-singular model for $C)$, $\alpha'_i \cap \alpha'_j = \emptyset$ when $i > 2g$, $j \leq 2g$ or $i > 2g$, $j > 2g$, $|j-i| \neq 0$, g_C-g and for $2g < i \leq 2g+(g_C-g)$ α'_i intersects $\alpha'_{i+(g_C-g)}$ transversally and in one point. Evidently $C' - \bigcup_{i=1}^{2g_C} \alpha'_i$ is connected and we can find disjoint smooth paths δ^{ℓ}_i, $i = 1,2,3$, on $C' - \bigcup_{i=1}^{2g_C} \alpha'_i - \hat{d}_p$ such that δ^{ℓ}_i connects $\tilde{a}^{\ell}_{i\sigma(i)}$ with $\tilde{a}^{\ell}_{\sigma(i)i}$, where $\sigma\colon (1,2,3) \longrightarrow (1,2,3)$ is defined by $\sigma(1) = 2$, $\sigma(2) = 3$, $\sigma(3) = 1$, and $\delta^{\ell}_i \cap \partial C' = (\tilde{a}^{\ell}_{i\sigma(i)} \cup \tilde{a}^{\ell}_{\sigma(i)i})$.

Let $\beta^{\ell}_i = (\delta^{\ell}_i \bigcup_{\psi^{\ell}_i|\tilde{a}^{\ell}_{i\sigma(i)}} \gamma^{\ell}_{i\sigma(i)}) \bigcup_{\psi^{\ell}_{\sigma(i)}|\tilde{a}^{\ell}_{\sigma(i)i}} \gamma^{\ell}_{\sigma(i)i}$.

Take $u_o \in C' - \bigcup_{j=1}^{2g_C} \alpha'_j - \bigcup_{\ell=1}^{\nu} \bigcup_{i=1}^{3} \delta^{\ell}_i$ and connect u_o with all α'_j, $j = 1,\cdots,2g_C$, and δ^{ℓ}_i, $i = 1,2,3$, $\ell = 1,2,\cdots,\nu$, by disjoint smooth paths t_j, $j = 1,2,\cdots,g$, $2g+1,\cdots,2g+(g_C-g)$, t^{ℓ}_i, $i = 1,2,3$, $\ell = 1,2,\cdots,\nu$, where $\mathrm{int}(t_j)$, $\mathrm{int}(t^{\ell}_i) \subset C' - \bigcup_{j=1}^{2g_C} \alpha'_j - \bigcup_{\ell=1}^{\nu} \bigcup_{i=1}^{3} \delta^{\ell}_i$, the end-point of t_j is $\alpha'_j \cap \alpha'_{j+g}$ when $j \leq g$ and it is $\alpha'_j \cap \alpha'_{j+(g_C-g)}$ when $j \geq 2g+1$, the end-point of t^{ℓ}_i belongs to δ^{ℓ}_i. We obtain a bouquet $\hat{\mathcal{B}}$ of circles on $\hat{C}$ such that all elements of $\hat{\mathcal{B}}$ are "very close" to $t_j \cup \alpha_j$, $j \in (1,\cdots,g) \cup (2g+1,\cdots,2g+1+g_C-g)$

or $t_j \cup \alpha_{j+g}$ $(j \in (1,2,\cdots,g)$, or $t_j \cup \alpha_{j+g_C-g}(j \in (2g+1,\cdots,2g+g_C-g))$, or to $t_i^{\ell} \cup \beta_i^{\ell}$, $i = 1,2,3$, $\ell = 1,2,\cdots,\nu$. It is easy to verify that there exists a deformation retraction of $\hat{C}^{\cdot}$ on $\hat{\mathscr{B}}$ and we can consider $\hat{C}^{\cdot}$ as a regular neighborhood of $\hat{\mathscr{B}}$ in $\hat{C}^{\cdot}$. Thus we can consider $T\hat{C}^{\cdot}$ as a regular neighborhood of $\hat{\mathscr{B}}$ in $\hat{A}$. Using $\eta_o: \bar{C} \longrightarrow \hat{C}$ we define $\mathscr{B} = \eta_o^{-1}(\hat{\mathscr{B}})$ and we see that we can consider $T\bar{C}^{\cdot}$ as a regular neighborhood of $\mathscr{B}$ in $\bar{\bar{V}}$.

We will use now the following result of R. Mandelbaum (see [5]).

<u>Theorem</u> (<u>R. Mandelbaum</u>). Let M_1, M_2 be compact differential 4-manifolds, M_1 is simply-connected, $\mathscr{B} = \bigvee_{k=1}^{n} S_k^1$ be a bouquet of n circles, $i_j\colon \ \mathscr{B} \longrightarrow \operatorname{int}(M_j)$, $j = 1,2$, be "smooth" embeddings, $\mathscr{B}_j = i_j(\mathscr{B})$, $i_j'\colon \mathscr{B}_j \longrightarrow \mathscr{B}$ be a map inverse to $i_j\colon \mathscr{B} \rightarrow \mathscr{B}_j (\subset M_j)$, $\tau_j\colon T\mathscr{B}_j \longrightarrow \mathscr{B}_j$ be a regular neighborhood of $\mathscr{B}_j$ in M_j, $\underline{\tau}_j = \tau_j|\partial T\mathscr{B}_j$. Suppose that a diffeomorphism $\eta\colon \partial T\mathscr{B}_1 \longrightarrow \partial T\mathscr{B}_2$ is given which satisfies the condition: $\underline{\tau}_2 \cdot \eta = i_2 i_1' \underline{\tau}_1$. Let S_{k2}^1, $k = 1,2,\cdots,n$, be disjoint smooth circles in M_2 such that each S_{k2}^1 is isotopically equivalent to $i_2(S_k^1)$. Then $(\overline{M_1 - T\mathscr{B}_1}) \cup_\eta (\overline{M_2 - T\mathscr{B}_2})$ is diffeomorphic to $M_1 \# \tilde{M}_2$, where $\tilde{M}_2$ is obtained from M_2 by surgeries along S_{k2}^1, $k = 1,2,\cdots,n$.

In our situation this theorem gives the following: Let $\tilde{\alpha}_j$, $j = 1,2,\cdots,2g_C$, $\tilde{\beta}_i^{\ell}$, $i = 1,2,3$, $\ell = 1,2,\cdots,\nu$, be smooth circles in A which are isotopically equivalent to α_j', $j = 1,2,\cdots,2g_C$, β_i^{ℓ}, $i = 1,2,3$, $\ell = 1,2,\cdots,\nu$, correspondingly.

Then $\overline{(\bar{\bar{V}}-T\bar{C}^{\cdot})} \cup_{\eta^{\cdot}} \overline{(\hat{A}-T\hat{C}^{\cdot})}$ is diffeomorphic to $\bar{\bar{V}} \# \tilde{A}$, where $\tilde{A}$ is obtained from $\hat{A}$ by surgeries along $\tilde{\alpha}_j$, $j = 1,2,\cdots,2g_C$, β_i^{ℓ}, $i = 1,2,3$, $\ell = 1,2,\cdots,\nu$.

Let $\tilde{A}_I$ be a 4-manifold obtained from $\hat{A}$ by surgeries along β_i^{ℓ}, $i = 1,2$, $\ell = 1,2,\cdots,\nu$. We need the following

Lemma 3. Let M_1, M_2 be two compact differentiable 4-manifolds, M_1 be simply-connected, $t_i\colon D^3 \times S^1 \longrightarrow \mathrm{int}(M_1)$, $t_i'\colon S^2 \times D^2 \longrightarrow \mathrm{int}(M_2)$, $i = 1,2,3$, be smooth embeddings, $T_i = t_i(D^3 \times S^1)$, $T_i' = t_i'(S^2 \times D^2)$, $T_i \cap T_j = \emptyset$, $T_i' \cap T_j' = \emptyset$ for $i \neq j$, $i,j = 1,2,3$, $T = \bigcup_{i=1}^{3} T_i$, $T' = \bigcup_{i=1}^{3} T_i'$, $a_{12}, a_{13} \in \partial T_1$, $a_{12} \neq a_{13}$, $a_{21} \in \partial T_2$, $a_{31} \in \partial T_3$, $a_{12}', a_{13}' \in \partial T_1'$, $a_{12}' \neq a_{13}'$, $a_{21}' \in \partial T_2'$, $a_{31}' \in \partial T_3'$, γ_2, γ_3 (corresp. γ_2', γ_3') be smooth disjoint paths in M_1 (corresp. M_2) such that for $j = 2,3$ $\mathrm{int}\,\gamma_j \subset (\mathrm{int}\,M_1) - T$ (corresp. $\mathrm{int}\,\gamma_j' \subset (\mathrm{int}\,M_2) - T'$) and γ_j (corresp. γ_j') connects a_{1j} with a_{j1} (corresp. a_{1j}' with a_{j1}'). Let $\psi_i\colon \partial T_i \longrightarrow \partial T_i'$, $i = 1,2,3$ be diffeomorphisms respecting S^2-bundle structures over S^1 on $\partial T_i, \partial T_i'$ (which come from diffeomorphisms $t_i|\partial(D^3 \times S^1) = S^2 \times S^1\colon S^2 \times S^1 \longrightarrow \partial T_i$, $t_i'|\partial(S^2 \times D^2) = S^2 \times S^1\colon S^2 \times S^1 \longrightarrow \partial T_i'$ and projection $S^2 \times S^1 \longrightarrow S^1$) and such that for $j = 1,2$ $a_{1j}' = \psi_1(a_{1j})$, $a_{j1}' = \psi_j(a_{j1})$.

Define $\psi\colon T \longrightarrow T'$ by $\psi|_{T_i} = \psi_i$, and let $M = \overline{M_1 - T} \cup_{\psi} \overline{M_2 - T'}$ and $\tilde{M}$ be a 4-manifold obtained from M by surgeries along $s_2 = \gamma_2 \cup_{\psi|\partial\gamma_2} \gamma_2'$ and $s_3 = \gamma_3 \cup_{\psi|\partial\gamma_3} \gamma_3'$.

Let $(S^2xD^2)_i$, $i = 1,2,3$ be three different copies of S^2xD^2, $(S^2xS^1)_i = \partial(S^2xD^2)_i$.

Then there exist diffeomorphisms of S^2-bundles over S^1 $\varphi_1: \partial T_i' \longrightarrow (S^2xS^1)_i$, $i = 1,2,3$, such that if

$$M_2(\varphi) = ((((M_2-T')\cup_{\varphi_1}(S^2xD^2)_1)\cup_{\varphi_2}(S^2xD^2)_2)\cup_{\varphi_3}(S^2xD^2)_3)$$

then $\widetilde{M}$ is diffeomorphic to $M_1 \# M_2(\varphi)$.

Proof of Lemma 3.

Because M_1 is simply-connected we can find 4-disks D_i^4, $i = 1,2,3$, embedded in M_1, such that $\mathrm{int}(D_i^4)(i = 1,2,3)$ contains T_i and if $\widetilde{Y}_j$, $j = 2,3$, is the part of Y_j lying on $M_1-\bigcup_{i=1}^{3}D_i$ then $\widetilde{Y}_j$ intersects D_1^4 and D_j^4 transversally and each of them only in one point. Consider D_i^4, $=1,2,3$, as embedded in a 4-sphere S_i^4 and let $\widetilde{T}_i = S_i^4-T_i$. We can consider T_i as a tubular neighborhood of a circle embedded in S_i^4 and we get diffeomorphisms $e_i: \widetilde{T}_i \longrightarrow (S^2xD^2)_i$. We have also that if $e_i': \partial\widetilde{T}_i \longrightarrow (S^2xS^1)_i$ is the corresponding diffeomorphism of boundaries(we use $\partial T_i = \partial\widetilde{T}_i$) then e_i' is a diffeomorphism of S^2-bundles $\partial T_i \longrightarrow S^1$ and $(S^2xS^1)\xrightarrow{pr_i} S^1$ where $\partial T_i \longrightarrow S^1$ came from $\partial T_i = \partial t_i(D^3xS^1) = t_i\partial(D^3xS^1) = t_i(S^2xS^1)$ and projection $S^2xS^1 \longrightarrow S^1$. Define $\varphi_i = e_i'\cdot(\psi_i)^{-1}$ and $M_2(\varphi)$ as in the statement of the Lemma.

Let $\widetilde{D}_i^4 = e_i(\overline{S_i^4 - D_i^4})$. In an evident way we consider $\widetilde{D}_i^4$, $i = 1,2,3$, as subspaces in $M_2(\varphi)$. We see that we can identify M with $\overline{M_1 - \bigcup_{i=1}^{3} D_i^4} \cup_f \overline{(M_2(\varphi) - \bigcup_{i=1}^{3} \widetilde{D}_i^4}$ where f is defined by $f|\partial D_i^4 = f_i : \partial D_i^4 \longrightarrow \partial \widetilde{D}_i^4$ and $f_i = e_i|\partial D_i^4$. Identify $\overline{M_1 - \bigcup_{i=1}^{3} D_i^4}$ and $\overline{M_2(\varphi) - \bigcup_{i=1}^{3} \widetilde{D}_i^4}$ with the corresponding subspaces of M.

Taking a ball $D^4 \subset M_1$, with $\mathrm{int}(D^4) \supset [(\bigcup_{i=1}^{3} D_i^4) \cup \widetilde{\gamma}_2 \cup \widetilde{\gamma}_3]$ we reduce the proof to the case when M_1 is diffeomorphic to S^4 and we have to show that $\widetilde{M}$ is diffeomorphic to $M_2(\varphi)$.

Let $\overline{M} = \overline{(M_1 - D_1^4)} \cup_{f_1} \overline{(M_2(\varphi) - \widetilde{D}_1^4}$. We see that we can consider M as a 4-manifold obtained from $\overline{M}$ by surgeries along two 0-dimensional spheres, say S_2^o, S_3^o, embedded in $\overline{M}$, and consider $\widetilde{M}$ as obtained from M by surgeries along two 1-dimensional spheres s_2, s_3 appearing on M in the following way: If $S_j^o = \{b_{j1}, b_{j2}\}$, $j = 2,3$, $b_{j1}, b_{j2} \in \overline{M}$ then there are smooth disjoint paths $\overline{\gamma}_2, \overline{\gamma}_3$ on $\overline{M}$ such that $\overline{\gamma}_j$ connects b_{j1} with b_{j2}, $j = 2,3$, and s_2, s_3 on M are obtained canonically from $\overline{\gamma}_2$ and $\overline{\gamma}_3$. These considerations immediately show that $\widetilde{M}$ is diffeomorphic to $M_2(\varphi)$. (We use that if $a,b \in S^4$ and $\overline{S}$ is obtained from S^4 by surgery along $S^o = \{a,b\}$ then $\overline{S}$ is diffeomorphic to $S^1 \times S^3$ and if $\widetilde{S}$ is obtained from $\overline{S}$ by surgery along some $S^1 \times c$, $c \in S^3$, then $\widetilde{S}$ is diffeomorphic to S^4).

Lemma 3 is proved.

We return to the proof of Statement (*). Lemma 3 shows that $\tilde{A}_I$ is diffeomorphic to $A' \# \nu P$. Let $\tilde{A}_{II}$ be a 4-manifold obtained from $\tilde{A}_I$ by surgeries along α_i', $i = 1,2,\cdots,2g$ (we identify α_i' with their images on $\tilde{A}_I$). From the construction of A' we know that $A' \approx A \# \rho Q$, that is, $\tilde{A}_I$ is diffeomorphic to $A \# \rho Q \# \nu P$. Let α_i'', $i = 1,2,\cdots,2g$, be the images of α_i', $i = 1,2,\cdots,2g$, in A and A_{II} be a 4-manifold obtained from A by surgeries along α_i'', $i = 1,2,\cdots,2g$. Clearly $\tilde{A}_{II}$ is diffeomorphic to $A_{II} \# \rho Q \# \nu P$. But from Lemma 2 applied inductively we get that A_{II} is an S^2-bundle over S^2. Because $\rho \neq 0$ we have from a result of Wall (see [8]) that $\tilde{A}_{II}$ is diffeomorphic to $(P \# Q) \# \rho Q \# \nu P$. Let α_i'', $i = 2g+1,\cdots,2g_C$, (corresp. α_ℓ'', $\ell = 2g_C+1,\cdots,2g_C+\nu$) be the images of α_i', $i = 2g+1,\cdots,2g_C$, (corresp. $\beta_3^{\ell-2g_C}$, $\ell = 2g_C+1,\cdots,2g_C+\nu$) in $\tilde{A}_{II}$. $\tilde{A}$ is a 4-manifold obtained from $\tilde{A}_{II}$ by surgeries along α_i'', $i = 2g+1,\cdots,2g_C+\nu$. Because $\tilde{A}_{II}$ is simply connected and has an odd intersection form we have from results of Wall (see [8]) that $\tilde{A}$ is diffeomorphic to $(P \# Q) \# \rho Q \# \nu P \# [2(g_C-g)+\nu] (P \# Q)$. From the classical Hurwitz formula we have: $2g_C-2 = 2(2g-2)+\rho$, that is, $2(g_C-g) = 2g-2+\rho$. We see that $\tilde{A}$ is diffeomorphic to

$$(P \# Q) \# \rho Q \# \nu P \# (2g-2+\rho+\nu)(P \# Q) \approx$$

$$(2\nu+\rho+2g-1)P \# (\nu+2\rho+2g-1)Q.$$

This finishes the proof of Statement (*) and also the proof of Theorem 2. Q.E.D.

§3. Comparison of topology of simply-connected projective surfaces of degree n and non-singular hypersurfaces of degree n in $\mathbb{C}P^3$.

We shall use an old classical result of projective algebraic geometry. Because we need it in slightly modified form, we shall give the main parts of its proof in the Appendix to Part I (see p. 99). The result is the following:

Theorem 3. Let V be an irreducible algebraic surface in $\mathbb{C}P^N$ with only isolated singular points, say $a_1, \cdots, a_q$. Suppose that for any $i = 1,2,\cdots,q$ the dimension of the Zariski tangent space of V at a_i is equal to three. Then for generic projection $\pi: V \longrightarrow \mathbb{C}P^3$ we have the following:

(1) There exist open neighborhoods U_i, $i = 1,2,\cdots,q$, of $\pi(a_i)$ in $\pi(V)$ such that $\forall\, i = 1,2,\cdots,q$

$\pi|_{\pi^{-1}(U_i)}: \pi^{-1}(U_i) \longrightarrow U_i$ is biregular.

(2) $\pi(V) - \bigcup_{i=1}^{q} \pi(a_i)$ has only ordinary singular points.

(3) If $N \geq 5$, V is not contained in some proper projective subspace of $\mathbb{C}P^N$ and $V \longrightarrow \mathbb{C}P^N$ is not the Veronese embedding of $\mathbb{C}P^2$ in $\mathbb{C}P^5$, corresponding to monomials of degree two, then the singular locus of $\pi(V) - \bigcup_{i=1}^{q} \pi(a_i)$ is an irreducible algebraic curve $S_\pi(V)$ in $\pi(V)$ and $\pi^{-1}(S_\pi(V))$ is irreducible.

(4) If $N > 3$ and V is not contained in some 3-dimensional projective subspace of $\mathbb{C}P^N$ then $\pi(V) - \bigcup_{i=1}^{q} \pi(a_i)$ has pinch-points.

<u>Theorem 4</u>. Let V_n be a projective algebraic surface of degree n embedded in $\mathbb{C}P^N$, $N \geq 5$, such that V_n is not in a proper projective subspace of $\mathbb{C}P^N$. Suppose that V_n is non-singular or has as singularities only rational double-points. Let $h: \bar{V}_n \longrightarrow V_n$ be the minimal desingularization of V_n. Let X_n be a non-singular hypersurface of degree n in $\mathbb{C}P^3$.

Suppose $\pi_1(\bar{V}_n) = 0$. Then

0) $b_+(\bar{V}_n) < b_+(X_n)$, $b_-(\bar{V}_n) < b_-(X_n)$

(in particular, $b_2(\bar{V}_n) < b_2(X_n)$) ;

0_a) $\mathrm{ind}(\bar{V}_n) > \mathrm{ind}(X_n)$;

1) $\bar{V}_n \# [b_+(X_n)-b_+(\bar{V}_n)+1]P \# [b_-(X_n)-b_-(\bar{V}_n)]Q$

is diffeomorphic to $X_n \# P$;

2) $\bar{V}_n \# [b_+(X_n)-b_+(\bar{V}_n)+1]P \# [b_-(X_n)-b_-(\bar{V}_n)]Q$

is diffeomorphic to $[b_+(X_n)+1]P \# [b_-(X_n)]Q$.

More precisely, let $\pi: V_n \longrightarrow \mathbb{C}P^3$ be a generic projection of V_n in $\mathbb{C}P^3$, $V_n' = \pi(V_n)$, $\nu = \nu(V_n')$ (corresp. $\rho = \rho(V_n')$) be the number of triplanar points (corresp. pinch-points) of V_n', $S = S(V_n')$ be the locus of ordinary singularities of V_n', $d = d(S(V_n'))$ be the degree of S, and in the case when S is irreducible (that is, $V_n \neq \mathbb{C}P^2[2]$ where $\mathbb{C}P^2[2]$ is the image of Veronese embedding $\mathbb{C}P^2 \longrightarrow \mathbb{C}P^5$ corresponding to monomials of degree two, see Theorem 3 (3)), let $g = g(S(V_n'))$ be the genus of (non-singular model of) S. If $V_n = \mathbb{C}P^2[2] (\subset \mathbb{C}P^5)$, let $g = g(S(V_n')) = -2$.

Then

0)' i) $b_+(X_n) - b_+(\overline{V}_n) =$

$$= \frac{n}{3}(n^2-6n+11)-1-b_+(\overline{V}_n) =$$

$$= -\nu + \frac{\rho}{2} + d(n-4) =$$

$$= 2\nu + \rho + 2g-2,$$

ii) $b_-(X_n) - b_-(V_n) =$

$$= \frac{n-1}{3}(2n^2-4n+3) - b_-(V_n) =$$

$$= -2\nu + \frac{3}{2}\rho + d(2n-4) =$$

$$= \nu + 2\rho + dn + 2g - 2;$$

$0_a)'$ $\mathrm{Ind}(\overline{V}_n) - \mathrm{Ind}(X_n) = \frac{2}{3}(dn-1) + \frac{4}{3}d + \frac{7}{6}\rho + \frac{2}{3}g$

1)'-2)' $\overline{V}_n \# [-\nu+\frac{\rho}{2}+d(n-4)+1]P \# [-2\nu+\frac{3}{2}\rho+d(2n-4)]Q =$

$$= \overline{V}_n \# [2\nu+\rho+2g-1]P \# [\nu+2\rho+dn+2g-2]Q \approx$$

$$\approx X_n \# P \approx k_nP \# \ell_nQ,$$

where $k_n = \frac{n}{3}(n^2-6n+11)$, $\ell_n = \frac{n-1}{3}(2n^2-4n+3)$.

<u>Proof</u>. If $V = \mathbb{C}P^2[2]$ we can directly verify that all statements of Theorem 4 are true. Thus suppose that $V \neq \mathbb{C}P^2[2]$. Consider generic projection $\pi: V_n \longrightarrow \mathbb{C}P^3$ and let $\pi': V_n \longrightarrow V_n'$ be the corresponding map of V_n to its image. Let X_n be a non-singular hypersurface of degree n in $\mathbb{C}P^3$ with the properties: X_n does not contain the images of rational double points of V_n and X_n is transversal to $S = S(V_n')$.

It is well known that the complex dimension of the Zariski tangent space for a rational double point is equal three ([9]). Now Theorem 3 shows that we can apply to $V_n', X_n \subset \mathbb{C}P^3$ Theorem 2.

Evidently, $b = S.X_n = dn \neq 0$ and we obtain from Theorem 2, 1) that $X_n \# P$ is diffeomorphic to

$$\overline{V}_n \# [2\nu+\rho+2g-1]P \# [\nu+2\rho+dn+2g-2]Q.$$

Let E' be a generic plane section of V_n', $E = \pi'^{-1}(E')$, $C = \pi'^{-1}(S)$, K_V be a canonical divisor on V (that is, $K_V = h(K_{\overline{V}})$ where $K_{\overline{V}}$ is a canonical divisor on $\overline{V}$), $\tilde{\pi}': V_n' \longrightarrow \mathbb{C}P^2$ be a generic projection of V_n' on $\mathbb{C}P^2$, $\tilde{\pi} = \tilde{\pi}'\pi'$, D be the ramification locus of $\tilde{\pi}$ in V_n and $F = 0$ be the equation of V_n' in $\mathbb{C}P^3$. Considering a partial derivative of F corresponding to $\tilde{\pi}': V_n' \longrightarrow \mathbb{C}P^2$ we can easily see that $(n-1)E \equiv C+D$ ($\equiv$ means here "is linearly equivalent"). Because $K_{\mathbb{C}P^2} \equiv -3\ell$ where ℓ is a projective line in $\mathbb{C}P^2$ we have $K_V \equiv -3E+D$. Thus $D \equiv K_V+3E$ and

$$C \equiv (n-1)E - D \equiv (n-4)E - K_V.$$

Let g_C (corresp. g_E) be the genus of (the non-singular model of C (corresp. E). Because C has 3ν ordinary double points, E' is a generic projection of E in $\mathbb{C}P^2$ and E' is of degree n and has d ordinary double points we have

$$2g_C-2 = (K_V+C)C - 6\nu = (n-4)E((n-4)E-K_V) - 6\nu =$$

$$= (n-4)^2 n - (n-4)E.K_V - 6\nu,$$

$$(K_V+E)E = 2g_E-2 = n(n-3) - 2d$$

and

$$K_V.E = n(n-4)-2d$$

$$2g_C-2 = (n-4)^2 n - (n-4)[n(n-4)-2d]-6\nu = 2d(n-4)-6\nu .$$

Let $\tilde{C},\tilde{S}$ be the normalizations of C and S and p: $\tilde{C} \longrightarrow \tilde{S}$ be the canonical map corresponding to $\pi'|_C: C \longrightarrow S$. Because p is of degree two and has ρ branch-points we have

$$2g_C-2 = 2(2g-2) + \rho \quad \text{and}$$

$$2g-2 = \frac{1}{2}(2g_C-2) - \frac{\rho}{2} = d(n-4) - 3\nu - \frac{\rho}{2}.$$

We see that $2\nu+\rho+2g-1 = -\nu+\frac{\rho}{2}+d(n-4)+1$ and

$$\nu+2\rho+dn+2g-2 = -2\nu+\frac{3}{2}\rho+d(2n-4).$$

Note also that in [2] it is proven that $X_n \# P$ is diffeomorphic to $k_n P \# \ell_n Q$ where $k_n = \frac{n}{3}(n^2-6n+11)$, $\ell_n = \frac{n-1}{3}(2n^2-4n+3)$. We see that the statements 0'), 1')-2') of our theorem are true. The statement $O_a)'$ can be verified now by direct calculation (using the formula $2g-2 = d(n-4)-3\nu-\frac{\rho}{2}$).

Now we have only to verify the statement O).

But it immediately follows from O') using $\rho \neq 0$ (Theorem 3 (4)) and $\rho \equiv 0(2)$. (If $b_+(X_n) = b_+(\overline{V}_n)$ we have $g = 0$, $\rho = 2$, $\nu = 0$. Then $d(n-4) = -1$ and $d = 1$, $n = 3$. But for a surface V_3 we can always find some $\mathbb{C}P^4$ in $\mathbb{C}P^N$ which contains V_3). Q.E.D.

§4. Simply-connected algebraic surfaces of general type.

Theorem 5. Let V be a simply-connected non-singular (complex) algebraic surface of general type, V_{min} be the minimal model of V, $c = c(V) = K^2_{V_{min}}$ (self-intersection of the canonical class), $\tilde{c} = K^2_V$, $p_g = p_g(V)$ (geometrical genus of V), $b_+ = b_+(V)$, $b_- = b_-(V)$, and $k(X)$, $\ell(X)$, $K(X)$, $L(X)$, $\tilde{K}(X)$, $\tilde{L}(X) \in \mathbb{Q}[X]$ be polynomials of degree three defined as follows:

$$\tilde{K}(X) = \frac{X}{3}(X^2-6X+11); \quad \tilde{L}(X) = \frac{X-1}{3}(2X^2-4X+3);$$

$$K(X) = \tilde{K}(9(5X+4))-X; \quad L(X) = \tilde{L}(9(5X+4));$$

$$k(X) = K(2X+1), \quad \ell(X) = L(2X+1) .$$

Let m be a positive integer and

$$\tilde{V}_m = V \# [\tilde{K}(m^2c)-2p_g-1]P \# [\max(0,\tilde{L}(m^2c)-10p_g-9+\tilde{c})]Q,$$

$$\tilde{V} = \begin{cases} \tilde{V}_3 & \text{if } c \geq 6 \text{ or } c \geq 3 \text{ and } p_g \geq 4, \\ \tilde{V}_4 & \text{if } c = 2 \text{ or } c = 3,4,5 \text{ and } p_g \leq 3, \\ \tilde{V}_5 & \text{if } c = 1. \end{cases}$$

Then

(1) $\tilde{V}$ is completely decomposable;

(2) $V \# [K(b_+)]P \# [\max(0,L(b_+)-b_-)]Q =$

$= V \# [k(p_g)]P \# [\max(0,\ell(p_g)-b_-)]Q$

is completely decomposable.

[Remark. Note that

$$K(b_+) = 30375b_+^3 + 68850b_+^2 + 52004b_+ + 13092;$$
$$L(b_+) = 60750b_+^3 + 141750b_+^2 + 110265b_+ + 28595;$$
$$k(p_g) = 243000p_g^3 + 251100p_g^2 + 211738p_g + 164321;$$
$$\ell(p_g) = 486000p_g^3 + 1296000p_g^2 + 1152050p_g + 341360.]$$

Proof. Using $V \approx V_{min} \# aQ$ where $a = c-\tilde{c} = b_- - b_-(V_{min}) \geq 0$ and $b_+ = 2p_g+1$ (Hodge Index Theorem) it is easy to verify that it is enough to prove our Theorem only for the case $V = V_{min}$. Thus assume $V = V_{min}$.

Let $m \geq 0$ be an integer such that the linear system $|mK_V|$ has no base points, the regular map $\mathcal{F}_m : V \longrightarrow \mathbb{C}P^{N(m)}$ corresponding to $|mK_V|$ is birational and $\mathcal{F}_m(V)$ has as singularities only rational double-points. Let $\underline{V} = \mathcal{F}_m(V)$ and $h: V \longrightarrow \underline{V}$ be the map corresponding to $\mathcal{F}_m$. We see that $h: V \longrightarrow \underline{V}$ is the minimal desingularization of $\underline{V}$ and we can apply to $\underline{V}$ the Theorem 4. Note that the projective degree of $\underline{V}$ is equal to $(mK.mK) = m^2c$ and $b_- = 10p_g+9-c$ (because $b_2(V) = b_++b_- = 2p_g+1+b_-$ (Hodge Index Theorem) and $1+p_g = \frac{1}{12}(c+2+b_2(V))$ (Noether Formula)).

We get from Theorem 4 that $\tilde{V}_m$ is completely decomposable. Now results of E. Bombieri[3] show that we can take $m = 3$

if $c \geq 6$ or $c \geq 3$ and $p_g \geq 4$, $m = 4$ if $c \geq 2$ and $m = 5$ in all cases. This proves the Statement (1) of Theorem 5.

The Statement (2) follows from (1). We have to remark only that $3^2c \leq 9(5b_+ +4)$ (because $c = 5b_+ +4-b_-$), $4^2c \leq 9(5b_+ +4)$ for $c \leq 5$ (because $b_+ \geq 1$ and $4^2c \leq 80$, $9(5b_+ +4) \geq 81$) and $5^2c \leq 9(5b_+ +4)$ for $c = 1$.

Q.E.D.

§5. Topological normalization of simply-connected algebraic surfaces.

Recall that we say that an algebraic complex function field R is topologically normal if there exists a non-singular model $V = V(R)$ of R such that V is almost completely decomposable (see [4]). Let R',R be two algebraic complex function fields of two variables. We say that R' is a satisfactorily cyclic extension of R if there exist non-singular models V' and V for R' and R correspondingly and a regular map $f: V' \longrightarrow V$ such that f is a ramified covering and the ramification locus of f in V is non-singular and linearly equivalent to $(\deg f).D$ where $D^2 > 0$ and $|D|$ is a linear system in V without base-points and fixed components. (This is a small modification of the corresponding definition in [4].)

We define $\pi_1(R) = \pi_1(V)$ where V is any non-singular model of R.

Definition 2. Let R be an algebraic complex function field of two variables with $\pi_1(R) = 0$. We shall say that R' is a topological normalization of R if R' is a satisfactorily cyclic extension of R and R' is topologically normal.

It was proven in [4] that for any R with $\pi_1(R) = 0$ there exists a topological normalization of R which is a quadratic extension of R. Now we shall give to this result more explicit form.

Theorem 6. Let R be an algebraic complex function field of two variables of general type, $\pi_1(R) = 0$, V be a non-singular model of R embedded in a projective space $\mathbb{C}P^N$ with homogeneous coordinates $(z_o,\cdots,z_N)$. Let $n = \deg V$, $m = \left[\frac{n}{\sqrt{3}}\right] + 1$ ($\left[\frac{n}{\sqrt{3}}\right]$ is the integer part of $\frac{n}{\sqrt{3}}$).

Suppose that V is not contained in the hyperplane given by $z_o = 0$. Let $\mathfrak{F}_{2m}(z_o:\ldots\ldots:z_N)$ be a homogeneous form of degree 2m of $z_o,\cdots,z_N$ such that the corresponding hypersurface section of V is non-singular. Let $f = \left(\frac{\mathfrak{F}_{2m}(z_o:\ldots:z_N)}{z_o^{2m}}\right)\Big|_V$ and $R' = R(\sqrt{f})$. Then R' is a topological normalization of R.

Proof. The theorem follows from Theorem 4, results of [4], [10], Sections 4 and 5, and from the following remarks:

a) If E_m is a hyperplane section of V of degree m then $2g(E_m)-2 = (K_V+mE_1)mE_1 = mK_VE_1+m^2n \geq m^2n+m$ (because $K_V\cdot E_1 > 0$).

b) For $m = \left[\frac{n}{\sqrt{3}}\right]+1$

$$m^2n + m + 2 \geq \frac{n}{3}(n^2-6n+11) - b_+.$$

Theorem 7. Let R be an algebraic complex function field of two variables of general type, $\pi_1(R) = 0$, $p_g = p_g(R)$ (geometrical genus) $P_2 = P_2(R)$(2-genus), $c = P_2-p_g-1$. Denote by

$$a = a(R) = \begin{cases} 6([3\sqrt{3}c]+1) & \text{if } c \geq 6 \text{ or } c \geq 3 \text{ and } p_g \geq 4 \\ 8([\frac{16c}{\sqrt{3}}]+1) & \text{if } c = 2 \text{ or } c = 3,4,5 \text{ and } p_g \leq 3 \\ 150 & \text{if } c = 1. \end{cases}$$

Let α_{2a}, β_a be two regular pluridifferentials of R of degree 2a and a correspondingly such that $\beta_a \neq 0$ and on some non-singular model V of R zero-divisor $(\alpha_{2a})_o$ of α_{2a} is a non-singular curve on V. Let $f = \frac{\alpha_{2a}}{\beta_a^2}$ and $R' = R(\sqrt{f})$. Then R' is a topological normalization of R.

Proof. The theorem follows from Theorem 5, results of [4], [10] and from the following remarks:

Let V be a minimal (non-singular) model of R, $m' = \frac{a}{2}$ and $E_{m'}$ be a non-singular element of $|im'K_V|$ where $i = 3$ if $c \geq 6$ or $c \geq 3$ and $p_g \geq 4$, $i = 4$ if $c = 2$ or $c = 3,4,5$ and $p_g \leq 3$ and $i = 5$ if $c = 1$. Let $n = i^2c$. Then $K_V^2 = P_2 - p_g - 1 = c$ (because $H^1(V, [2K_V]) = 0$ (see [11]). As in the proof of Theorem 6 we have $2g(E_{m'})-2 \geq (m')^2n+m'$ and $(m')^2n+m'+2 \geq \frac{n}{3}(n^2-6n+11)-b_+$.

Q.E.D.

APPENDIX TO PART I

GENERIC PROJECTIONS OF ALGEBRAIC SURFACES INTO $\mathbb{C}P^3$

§1. A theorem of F. Severi.

The following theorem was proved by F. Severi in 1901 (see [12]) in non-singular case. We need a slightly more general fact. Our proof is very close to Severi's arguments.

Theorem (F. Severi). Let V be an irreducible algebraic surface with only isolated singular points embedded in $\mathbb{C}P^5$ such that V is not contained in a proper projective subspace of $\mathbb{C}P^5$. Let K(V) be the variety of chords of V (that is, the algebraic closure in $\mathbb{C}P^5$ of the union of all projective lines in $\mathbb{C}P^5$ connecting two different points of V).

Then $4 \leq \dim_{\mathbb{C}} K(V) \leq 5$ and $\dim_{\mathbb{C}} K(V) = 4$ iff V is a projective cone over some algebraic curve or the given embedding $V \longrightarrow \mathbb{C}P^5$ coincides with the Veronese embedding of $\mathbb{C}P^2$ corresponding to monomials of degree two (that is, $V = \mathbb{C}P^2[2]$).

Proof. It is clear that K(V) is irreducible and $\dim_{\mathbb{C}} K(V) \leq 5$.

Suppose that $\dim_{\mathbb{C}} K(V) \leq 3$. Let E be a generic hyperplane section of V corresponding to some hyperplane $\mathbb{C}P^4_E \subset \mathbb{C}P^5$. Since $V \not\subset \mathbb{C}P^4_E$, we have $K(E) \neq K(V)$ (K(E) is the variety of chords of E). Hence $\dim_{\mathbb{C}} K(E) < 3$. Because E is non-singular, we obtain from $\dim_{\mathbb{C}} K(E) < 3$ that generic projection of E in $\mathbb{C}P^2$ is also non-singula

Denote by E' the image of this projection (clearly, E' is isomorphic to E). Let $[\ell]_{E'}$ be the restriction on E' of a complex line bundle on $\mathbb{C}P^2$ corresponding to the projective line in $\mathbb{C}P^2$. Then $\dim_{\mathbb{C}} H^0(E', \mathcal{O}[\ell]_{E'}) = 3$. It follows from this that E is contained in some $\mathbb{C}P^2 \subset \mathbb{C}P^4_E$. Hence V is contained in a proper projective subspace of $\mathbb{C}P^5$. Contradiction.

Now consider the case $\dim_{\mathbb{C}} K(V) = 4$. Let x be a generic point of K(V). Denote by Γ_x the cone in $\mathbb{C}P^5$ which is union of all "chords" of V passing through x. It is clear that $\dim_{\mathbb{C}} \Gamma_x \leq 2$. Suppose $\dim_{\mathbb{C}} \Gamma_x = 1$. Then using $\dim_{\mathbb{C}} K(X) = 4$ we have that a generic "chord" of V contains an infinite number of points of V, that is, $V = \mathbb{C}P^2$. Contradiction. Thus $\dim_{\mathbb{C}} \Gamma_x = 2$.

Consider a generic hyperplane $\mathbb{C}P^4_E$ of $\mathbb{C}P^5$ and let $E = V \cap \mathbb{C}P^4_E$. As above, $\dim_{\mathbb{C}} K(E) = 3$. Because $K(V)$, $K(V) \cap \mathbb{C}P^4_E$ and $K(E)$ are irreducible, $K(E) \subset K(V) \cap \mathbb{C}P^4_E$ and $\dim_{\mathbb{C}}(K(V) \cap \mathbb{C}P^4_E) = 3$, we have $K(E) = K(V) \cap \mathbb{C}P^4_E$. We can assume that x is generic on K(E). Let $\ell_1, \cdots, \ell_m$ be all the chords of E passing through x. It is clear that $\Gamma_x \cap \mathbb{C}P^4_E = \bigcup_{i=1}^{m} \ell_i$.

We can assume that x is a generic point on a generic "chord" of E. Suppose $m > 1$. Then for a generic point x of a generic "chord" of E there exists another "chord" of E passing through x.

Let a,b be two different generic points of E, and e be a generic point of the "chord" k(a,b) corresponding to a,b. From our supposition it follows that there exists another "chord" k(c,d) of E, where $c,d \in E$, $c \neq d$, such that $e \in k(c,d)$. Because e is generic we can assume that c is a generic point of E. Because generic projection of E in $\mathbb{C}P^2$ has as singularities only ordinary double points, $\dim K(E) = 3$ and a,b are generic on E we have: $k(a,b) \cap E = \{a,b\}$ and $c \notin k(a,b)$, $d \notin k(a,b)$.

Now let $\pi_a : E \longrightarrow \mathbb{C}P^3$ be projection of E in $\mathbb{C}P^3$ with the center a, $E' = \pi_a(E)$ (that is, the closure of $\pi_a(E-a)$). Let M be the two-plane in $\mathbb{C}P^4_E$ containing $k(a,b) \cup k(c,d)$, $M' = \pi_a(M)$, $b' = \pi_a(b)$, $c' = \pi_a(c)$, $d' = \pi_a(d)$.

Since b and c are generic on E, $c,d \notin k(a,b)$, $a \notin k(c,d)$, we have that b' and c' are generic on E', $b' \neq c'$, $b' \neq d'$, $c' \neq d'$. We have now that $M' = k(b',c')$ (which is a generic "chord" of E') meets E' in a third point.

Because E is not in a proper projective subspace of $\mathbb{C}P^4_E$, E' is not a plane curve in $\mathbb{C}P^3$. It easily follows from this that $\dim_{\mathbb{C}} K(E') = 3$, that is, $K(E') = \mathbb{C}P^3$. A generic projection of E' in $\mathbb{C}P^2$ has the following property: if E" is the image of E' then E" has as singularities only the images of the singular points of E' with all the branches as images of the branches of the singular points of E' and finite number of ordinary double points. [This can be proved by stability arguments in the singular case as well as in the non-singular case]. We get a contradiction with the facts: $K(E') = \mathbb{C}P^3$ and a generic chord of E contains three different points.

Thus we obtain m = 1. But this means that x is a non-singular point of Γ_x, that is, Γ_x is a plane in $\mathbb{C}P^5$. Let $C_x = \Gamma_x \cap V$. From $\Gamma_x \cap \mathbb{C}P^4_E = \ell_1$ and ℓ_1 is a generic "chord" of V, that is, it meets V only in two points (because ℓ_1 is also a generic "chord" of E) we see that C_x (which must be a curve) has projective degree two. The construction of Γ_x gives us a rational map of K(V) in the grassmanian of 2-planes of $\mathbb{C}P^5$. Let T be the image of this map. Since $\dim_{\mathbb{C}} K(V) = 4$ and $\dim_{\mathbb{C}} \Gamma_x = 2$ we have that $\dim_{\mathbb{C}} T \geq 2$. It is clear that for generic $x,y \in K(V)$ corresponding C_x, C_y are algebraically equivalent and let us consider now the maximal irreducible algebraic system $\{C_{(t)}, t \in T'\}$ of curves on V which contains C_x as a generic element (T' is a subvariety of the corresponding Chow variety). If for generic $x,y \in K(V)$ we would have $\Gamma_x \neq \Gamma_y$ and $C_x = C_y$ then C_x is a projective line. This contradicts to $\deg(C_x) = 2$.

But that means that the natural rational map $T \longrightarrow T'$ is of degree one and $\dim_{\mathbb{C}} T' \geq 2$.

Now let us consider the different possible cases.

1. Generic C_x contains some of the singular points of V. Because V has only a finite number of singular points and T' is irreducible we see that there exists a singular point a_o of V such that every $C_{(t)}$, $t \in T'$, contains it.

1a. Generic C_x is irreducible. Since $\dim_{\mathbb{C}} T' \geq 2$ then a generic point $a \in V$ is contained in infinitely many irreducible elements of $\{C_{(t)}\}$. Let C_1, C_2 be two of them, $C_2 \cap C_1 \ni a$, a is non-singular on V, and hence for any $C_{(t)}$ which is "close" to C_2 we have that $C_{(t)}$ intersects C_1 in some point a' which is "close" to a. That means that there exists a non-empty Zariski open subset U_1 in T' such that for any $t \in U_1$ corresponding $C_{(t)}$ intersects C_1 in some point $a(t) \neq a_o$. By the same reason we have analogous $U_2 \subset T'$ for C_2. C_1 and C_2 are of degree two and let Γ_1, Γ_2 be 2-planes containing C_1 and C_2 correspondingly. Since $a_o \in C_1 \cap C_2 \subset \Gamma_1 \cap \Gamma_2$ there exists a hyperplane $\mathbb{C}P^4_{12}$ in $\mathbb{C}P^5$ which contains Γ_1 and Γ_2. Let E_{12} be the corresponding hyperplane section of V. Evidently $E_{12} = C_1 + C_2 + D$, where D is some non-negative divisor (of Weil) in V. Because $V \not\subset E_{12} \subset \mathbb{C}P^4_{12}$ we have that there exists a non-empty Zariski open subset $U \subset T'$ such that for any $t \in U$ the corresponding $C_{(t)}$ is not in E_{12}. All this shows that we can choose such $C_3 \in \{C_{(t)}\}$ that C_3 is irreducible, different from C_1, C_2, not a component of D and intersects C_1 in some $a' \neq a_o$ and C_2 in some $b' \neq a_o$. We have $E_{12} \cap C_3 \supset a_o, a', b'$. If $a' = b'$ we have from $C_1 \neq C_2$ that intersection index of C_3 with $C_1 + C_2$ in a' is greater than one. In any case, we get $\mathbb{C}P^4_{12} \cdot C_3 \geq 3$. Contradiction ($\deg C_{(t)} = 2$).

1b. Generic C_x is reducible. Then each component of C_x must be a projective line in $\mathbb{C}P^5$. Suppose that there exists only a finite number of such lines which contain a_o. This means that all $C_{(t)}$ have a common line, say ℓ_1. Thus every chord of V intersects ℓ_1. Take generic $\mathbb{C}P^4_E \not\ni \ell_1$ and the corresponding E. We obtain that any chord of E is passing through some finite number of points of $E \cap \ell_1$ and thus it contains three different points. This contradicts $\dim_{\mathbb{C}} K(E) = 3$ and the fact that a generic projection of E on $\mathbb{C}P^2$ has as singularities only ordinary double points.

We get that there exist an infinite number of projective lines of $\mathbb{C}P^5$ in V containing a_o. The existence of the Chow variety of projective lines of $\mathbb{C}P^5$ which are contained in V gives us an irreducible 1-parametric algebraic system of such lines which are passing through a_o. We obtain that V is a cone over some algebraic curve.

2. Generic C_x does not contain a singular point of V.

2a. Generic C_x is reducible. As above we have an infinite number of projective lines of $\mathbb{C}P^5$ in V. The corresponding Chow variety shows that only a finite number of them can be "isolated", that is, not in some infinite irreducible algebraic system of such lines. If generic C_x would contain such an "isolated" line, then we would have that every $C_{(t)} \in \{C_t\}$ contains a certain line ℓ_1. This leads to a contradiction by the same arguments as above (Case 1b). We get that generic C_x is a union of two lines ℓ_{1x}, ℓ_{2x}, which both are in some irreducible algebraic systems. Thus $\ell_{1x}.\ell_{1x} \geq 0$, $\ell_{2x}.\ell_{2x} \geq 0$ and for any curve d in V, $C_x.d \geq 0$ (intersection numbers are defined because C_x lies in the non-singular part of V). Because C_x is a plane curve we have $\ell_{1x}.\ell_{2x} = 1$ and $(C_x^2)_V = \ell_{1x}^2+2\ell_{1x}\ell_{2x}+\ell_{2x}^2 \geq 2$. Take now a different generic C_y. If C_y contains ℓ_{1x} or ℓ_{2x} we get that all $C_{(t)} \in \{C_{(t)}\}$ have a common component which is impossible (as we saw above). Thus $C_y = \ell_{1y}+\ell_{2y}$, where ℓ_{1y}, ℓ_{2y} are different from ℓ_{1x}, ℓ_{2x} and not "isolated". If $C_x.C_y = 0$ then $\ell_{1x}.\ell_{1y} = \ell_{1x}\ell_{2y} = \ell_{2x}.\ell_{1y} = \ell_{2x}.\ell_{2y} = 0$. It is easy to see that these equalities give that ℓ_{1x} and ℓ_{2x} are in the same irreducible algebraic system as ℓ_{1y}. Thus $\ell_{1x}.\ell_{2x} = \ell_{1x}.\ell_{1y} = 0$. Contradiction.

We get $C_x.C_y \gneq 0$ and let $a \in C_x \cap C_y$. As above, we take a hyperplane containing C_x and C_y. Let $E_{x,y}$ be the corresponding hyperplane section. Because C_x and C_y have no common component we have that $E_{x,y} = C_x+C_y+D$ where D is some non-negative divisor (of A. Weil) on V. As we saw above, $C_x.D \geq 0$. We have

$$2 = E_{x.y}.C_x = C_x^2 + C_xC_y + D.C_x \geq 2+1 = 3.$$

Contradiction.

2b. Generic C_x is irreducible. Since $\dim T' \geq 2$ we have for a generic point $a \in V$ that there exists an infinite number of irreducible $C_{(t)}$ containing a. Take two different ones, say C_1 and C_2. We can assume also that C_1 and C_2 are in the non-singular part of V. We see that $C_1.C_2 > 0$, that is, $C_x^2 > 0$, and for any algebraic curve d on V, $C_x.D \geq 0$. As above, take a hyperplane section E_{12} of V containing C_1 and C_2. We have $E_{12} = C_1+C_2+D$, where D is a non-negative divisor (of Weil) on V, and

$$2 = E_{12}.C_x = 2C_x^2 + D.C_x.$$

We get $C_x^2 = 1$, $D.C_x = 0$. Now suppose that $D \neq 0$. E_{12} is connected and thus D either contains C_1 or C_2 as components or intersects with C_1 or C_2 in some points. In all cases we have a contradiction with $D.C_2 = 0$. Thus $D = 0$ and $E_{12} = C_1+C_2$. Now C_1 and C_2 are Cartier divisors on V and we can consider corresponding complex line bundles $[C_1]$, $[C_2]$ over V. Clearly $[E] = [C_1]+[C_2]$, where $[E]$ is the line bundle corresponding to hyperplane sections. We have the following exact sequences:

$$(1) \quad 0 \longrightarrow H^0(V, \mathcal{O}_V) \longrightarrow H^0(V, \mathcal{O}_V[C_1]) \longrightarrow H^0(C_1, \mathcal{O}_{C_1}[C_1]_{C_1})$$

$$(2) \quad 0 \longrightarrow H^0(V, \mathcal{O}_V[C_1]) \longrightarrow H^0(V, \mathcal{O}_V[C_1+C_2]) \longrightarrow H^0(C_2, \mathcal{O}_{C_2}[C_1+C_2]_{C_2}) .$$

C_1 and C_2 are both rational non-singular curves. From $C_1^2 = 1$, $(C_1+C_2)C_2 = 2$ we get

$$(3) \quad \dim_{\mathbb{C}} H^0(C_1, \mathcal{O}_{C_1}[C_1]_{C_1}) = 2, \quad \dim_{\mathbb{C}} H^0(C_2, \mathcal{O}_{C_2}[C_1+C_2]_{C_2}) = 3.$$

Because $\dim_{\mathbb{C}} H^0(V, \mathcal{O}_V) = 1$, we obtain from (1),(2),(3) and from the fact that V is not in a proper projective subspace of $\mathbb{C}P^5$

$$\dim_{\mathbb{C}} H^0(V, \mathcal{O}_V[C_1]) \leq 3$$

$$6 \leq \dim_{\mathbb{C}} H^0(V, \mathcal{O}_V[E]) = \dim_{\mathbb{C}} H^0(V, \mathcal{O}_V[C_1+C_2]) \leq$$

$$\leq \dim_{\mathbb{C}} H^0(V, \mathcal{O}_V[C_1]) + 3 \leq 6.$$

This shows that $\dim_{\mathbb{C}} H^0(V, \mathcal{O}_V[E]) = 6$ and $\dim_{\mathbb{C}} H^0(V, \mathcal{O}_V[C_1]) = 3$. Because $C_1^2 = 1$ and C_1 is irreducible we have that global cross-sections of $[C_1]$ have no common zero and we can define a regular map $f: V \longrightarrow \mathbb{C}P^2$ corresponding to $H^0(V, \mathcal{O}_V[C_1])$. It is easy to see that f is surjective. We can find a $\varphi \in H^0(V, \mathcal{O}_V[C_1])$ such that zero-divisor of φ, say C_1', is an irreducible algebraic curve in V. Because the degree of C_1' is two we have that C_1' is contained in some 2-plane of $\mathbb{C}P^5$. $C_1^2 = 1$ gives $C_1' \cap C_1 \neq \emptyset$ and we can find a hyperplane section E_1 of V containing C_1 and C_1',

$E_1 = C_1 + C_1' + D'$, $D' \geq 0$. From $E_1^2 = (C_1 + C_2)^2 = 4$, $E_1 . C_1 = E_1 . C_1' = 2$ we get $E_1 . D' = 0$, that is, $D' = 0$, $E_1 = C_1 + C_1'$. Let $[E']$ be the line bundle over $\mathbb{C}P^2$ corresponding to 2ℓ, ℓ is a line in $\mathbb{C}P^2$. From $E_1 = C_1 + C_1'$ we get that $[E] = f^*[E']$, and

$$H^o(V, \mathcal{O}_V[E] \supseteq f^* H^o(\mathbb{C}P^2, \mathcal{O}_{\mathbb{C}P^2}[E']).$$

But $\dim_{\mathbb{C}} H^o(V, \mathcal{O}_V[E]) = 6$ and also $\dim_{\mathbb{C}} H^o(\mathbb{C}P^2, \mathcal{O}_{\mathbb{C}P^2}[E']) = 6$. We have

(4) $$H^o(V, \mathcal{O}_V[E]) = f^* H^o(\mathbb{C}P^2; \mathcal{O}_{\mathbb{C}P^2}[E']).$$

Let $f_{[2]} : \mathbb{C}P^2 \longrightarrow \mathbb{C}P^5$ be the Veronese embedding of $\mathbb{C}P^2$ corresponding to $[E']$, $i : V \longrightarrow \mathbb{C}P^5$ our original embedding which (because of $\dim_{\mathbb{C}} H^o(V, \mathcal{O}_V[E] = 6)$ corresponds to $[E]$. $E = [f^*E']$ and (4) give us the following commutative diagram:

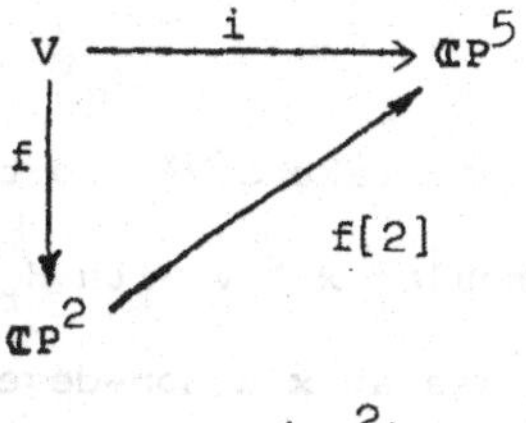

which shows that $V = f_{[2]}(\mathbb{C}P^2)$. Q.E.D.

§2. Duality theorem and Corollaries of it.

Definition. Let V be an irreducible algebraic variety embedded in $\mathbb{C}P^N$, $\dim_{\mathbb{C}} V = k$, τ_x for non-singular $x \in V$ be the k-dimensional projective subspace of $\mathbb{C}P^N$ tangent to V at x, $\mathbb{C}P^{N*}$ be the dual projective space for $\mathbb{C}P^N$ and V* be the algebraic closure in $\mathbb{C}P^{N*}$ of the set of all t such that the corresponding hyperplane H_t in $\mathbb{C}P^N$ contains τ_x for some non-singular $x \in V$. We call V* ($\subset \mathbb{C}P^{N*}$) the dual projective variety for V.

Duality Theorem. If V is as above, then

(1) V* is an irreducible proper subvariety of $\mathbb{C}P^{N*}$;

(2) there exists a proper subvariety S(V*) in V* such that for any $t \in V^*-S(V^*)$ the corresponding hyperplane H_t in $\mathbb{C}P^N$ contains τ_x for some non-singular $x \in V$, and for any such x the corresponding hyperplane H_x^* in $\mathbb{C}P^N$ is tangent to V* at t;

(3) V is dual to V*;

(4) in the case $\dim_{\mathbb{C}} V^* = N-1$ we can take S(V*) so that for any $t \in V^*-S(V^*)$ and for any non-singular $x \in V$ with $H_t \supset \tau_x$ the hyperplane section $E_t = H_t \cap V$ has at x a non-degenerate quadratic singular point.

Proof. (1) Let V_{sm} be the set of all non-singular points of V, $\Gamma' = \{(x,t) \in V_{sm} \times \mathbb{C}P^{N*}, H_t \supset \tau_x\}$, $f_1'\colon \Gamma' \longrightarrow V_{sm}$, $f_2'\colon \Gamma' \longrightarrow \mathbb{C}P^{N*}$ be the maps induced by canonical projections. It is easy to see that $f_1'\colon \Gamma' \longrightarrow V_{sm}$ is a fibre bundle over

V_{sm} with a typical fiber isomorphic to $\mathbb{C}P^{N-k-1}$ and that V^* is the algebraic closure of $f_2'(\Gamma')$ in $\mathbb{C}P^{N^*}$. We obtain that $\dim_{\mathbb{C}}\Gamma' = k+N-k-1 = N-1$ and that Γ' is irreducible. Hence $\dim_{\mathbb{C}}V^* \leq N-1$ and V^* is irreducible.

(2) Let $\tilde{V}$ be the algebraic closure of Γ' in $V \times \mathbb{C}P^{N^*}$, $f_1\colon \tilde{V} \longrightarrow V$, $f_2\colon \tilde{V} \longrightarrow V^*$ be the maps induced by f_1' and f_2'. Take a point $(v_o,t_o) \in \Gamma'$, such that t_o is nonsingular in V^* and df_2 is surjective in (v_o,t_o). Let $(x_1,\cdots,x_N)$ be some affine coordinate system with the center in v_o, $y_1,\cdots,y_N$ be the restrictions of $x_1,\cdots,x_N$ on V_x (in some neighborhood of v_o). We can assume that the affine coordinates are chosen so that there exists an open neighborhood U of v_o in V_{sm} such that $y_1,\cdots,y_k$ are local coordinates of V in U and $y_{k+1},\cdots,y_N$ are regular functions of $y_1,\cdots,y_k$ in U. We shall use capital letters $Y_{k+1},\cdots,Y_N$ instead of $y_{k+1},\cdots,y_N$.

If $v' \in U$ then any hyperplane of P^N passing through v' is defined by an equation: $a_o + \sum_{i=1}^{N} a_i x_i = 0$ where

$a_o + \sum_{i=1}^{N} a_i y_i(v') = 0$. This hyperplane is tangent to V^* at v' iff it contains the following vector (with the origin in v'):

$$(dy_1(v'),\cdots,dy_N(v')).$$

We have

$$\sum_{i=1}^{N} a_i dy_i(v') = 0$$

or

$$\sum_{j=1}^{k} a_j dy_j(v') + \sum_{i=k+1}^{N} a_i \sum_{j=1}^{k} \frac{\partial Y_i}{\partial y_j}(v')dy_j(v') = 0,$$

$$\sum_{j=1}^{k} (a_j + \sum_{i=k+1}^{N} a_i \frac{\partial Y_i}{\partial y_j}(v'))dy_j(v') = 0.$$

We obtain the following system of equations:

$$\sum_{i=1}^{N} a_i y_i(v') + a_o = 0,$$

$$a_j + \sum_{i=k+1}^{N} a_i \frac{\partial Y_i}{\partial y_j}(v') = 0, \quad 1 \leq j \leq k,$$

or

$$a_o = \sum_{i=k+1}^{N} a_i \Big[\sum_{j=1}^{k} \frac{\partial Y_i}{\partial y_j}(v')y_j(v') - Y_i(v') \Big] .$$

(1)
$$a_j = - \sum_{i=k+1}^{N} a_i \frac{\partial Y_i}{\partial y_j}(v'), \quad 1 \leq j \leq k.$$

These equations define an N-k-1-linear subspace $\mathfrak{P}(v')$ of $\mathbb{C}P^{N*}$ parametrizing all those hyperpla es of $\mathbb{C}P^N$ which are tangent to V at v'. It is clear that one of a_i, $i = k+1,\cdots,N$, is not zero at t_o. We can assume that $a_N(t_o) \neq 0$. Taking U smaller we can choose an open neighborhood U_N' of t_o in $\mathbb{C}P^{N*}$ such that $a_N \neq 0$ in U_N', $U_N' \cap V^*$ is a non-singular open subset of V^* and

df_2 is surjective in any point $(t',v') \in \Gamma'$ with $t' \in U_N' \cap V^*$, $v' \in U$. Take affine coordinates in U_N' defined as follows:

$$s_i = \frac{a_i}{a_N}, \quad i = 0,1,\cdots,N-1.$$

Now we write the equations {1} in the following form:

$$\{2\} \qquad \begin{aligned} s_o &= \sum_{i=k+1}^{N-1} s_i\Big(\sum_{j=1}^{k} \frac{\partial Y_i}{\partial y_j} y_j - Y_i\Big) + \sum_{j=1}^{k} \frac{\partial Y_N}{\partial y_j} y_j - Y_N; \\ s_j &= -\sum_{i=k+1}^{N-1} s_i \frac{\partial Y_i}{\partial y_j} - \frac{\partial Y_N}{\partial y_j}, \qquad j = 1,2,\cdots,k. \end{aligned}$$

Taking differentials we have

$$\{3\} \qquad \begin{aligned} ds_o &= \sum_{i=k+1}^{N-1}\Big(\sum_{j=1}^{k} \frac{\partial Y_i}{\partial y_j} y_j - Y_i\Big)ds_i + \sum_{\ell=1}^{k}\Big[\Big(\sum_{m=1}^{k}\ \sum_{i=k+1}^{N-1} \frac{\partial^2 Y_i}{\partial y_\ell \partial y_m} s_i y_m\Big) \\ &\qquad + \Big(\sum_{m=1}^{k} \frac{\partial^2 Y_N}{\partial y_\ell \partial y_m}\Big) y_m\Big]\, dy_\ell\ ; \\ ds_j &= -\sum_{i=k+1}^{N-1} \frac{\partial Y_i}{\partial y_j} ds_i - \sum_{\ell=1}^{k}\Big[\Big(\sum_{i=k+1}^{N-1} s_i \frac{\partial^2 Y_i}{\partial y_j \partial y_\ell}\Big) + \frac{\partial^2 Y_N}{\partial y_j \partial y_\ell}\Big] dy_\ell; \\ &\qquad j = 1,2,\cdots,k. \end{aligned}$$

Let $U' = U_N' \cap V^*$. Any tangent hyperplane in a generic point of U' is defined by a non-trivial linear combination of $ds_o,\cdots,ds_{N-1}$ equal to zero for any choice of $dy_1,\cdots,dy_k$, where $ds_o,\cdots,ds_{N-1}$ satisfy (3).

One such combination is

$$(4) \qquad ds_o + \sum_{j=1}^{k} y_j ds_j + \sum_{i=k+1}^{N-1} Y_i ds_i = 0 .$$

The corresponding hyperplane in $\mathbb{CP}^{N*}$ has the following equation:

$$(5) \quad s_o - s_o(t') + \sum_{j=1}^{k} y_j(v')(s_j - s_j(t')) + \sum_{i=k+1}^{N-1} Y_i(v')(s_i - s_i(t')) = 0.$$

Since

$$Y_N(v') = -s_o(t') - \sum_{j=1}^{k} y_j(v')s_j(t') - \sum_{i=k+1}^{N-1} Y_i(v')s_i(t'),$$

we can write (5) in the following form:

$$(6) \quad s_o + \sum_{j=1}^{k} y_j(v')s_j + \sum_{i=k+1}^{N-1} Y_i(v')s_i + Y_N(v') = 0.$$

The point in $\mathbb{CP}^N$ which corresponds to this hyperplane is v'.

There exists a non-empty Zariski open subset V_1^* in V^* such that V_1^* is non-singular, $V_1^* \subseteq f_2'(\Gamma')$ and for any $z \in f_2'^{-1}(V_1^*)$, $(df_2)_z$ is surjective. We see that we can take $S(V^*) = V^* - V_1^*$.

$(3)^*$) Let $k^* = \dim_{\mathbb{C}} V^*$, $\Gamma_1' = \{(x,t),\ x \in \mathbb{CP}^N,\ t \in V_1^*$ (V_1^* is the same as above), the hyperplane H_x^* in $\mathbb{CP}^{N*}$ corresponding to x contains the k^*-dimensional projective subspace τ_{t,V^*} of $\mathbb{CP}^{N*}$

* The proof of (3) which we give here is due to F. Catanese.

tangent to V^* at $t\}$. It follows from (2) that $f_2'^{-1}(V_1^*) \subseteq \Gamma_1'$. Let Γ_1 be the closure of Γ_1' in $\mathbb{C}P^N \times \mathbb{C}P^{N*}$. We obtain $\widetilde{V} \subseteq \Gamma_1$. Now the same arguments as in (1) show that Γ_1 is irreducible and $\dim_{\mathbb{C}} \Gamma_1 = N-1$. Because $\dim_{\mathbb{C}} \widetilde{V} = N-1$ we get $\widetilde{V} = \Gamma_1$. Evidently $pr_{\mathbb{C}P^N} \Gamma_1$ is the dual algebraic variety for V^*. Thus we proved that V is dual to V^*.

(4) Using the same notations as in (2) we can assume that the local equation of E_t in some neighborhood of x has the following form:

$$\sum_{j=1}^{k} s_j(t) y_j + \sum_{i=k+1}^{N-1} s_i(t) Y_i + Y_N = 0.$$

Denote $b_{j\ell} = \sum_{i=k+1}^{N-1} s_i(t) \frac{\partial^2 Y_i}{\partial y_j \partial y_\ell}(x) + \frac{\partial^2 Y_N}{\partial y_j \partial y_\ell}(x)$, $j,\ell = 1,2,\cdots,k$.

We must prove that $\mathrm{rk}\,\|b_{j\ell}\| = k$. Suppose that it is not true. Then there exist not all equal to zero constants $\bar{c}_1,\cdots,\bar{c}_k$ such that

$$\sum_{j=1}^{k} \bar{c}_j \Big(\sum_{i=k+1}^{N-1} s_i(t) \frac{\partial^2 Y_i}{\partial y_j \partial y_\ell}(x) + \frac{\partial^2 Y_N}{\partial y_j \partial y_\ell}(x) \Big) = 0 .$$

For $i = k+1,\cdots,N$ denote $\bar{c}_i = \sum_{j=1}^{k} \bar{c}_j \frac{\partial Y_i}{\partial y_j}(x)$.

Now we obtain from (3) that

$$(7) \qquad \sum_{i=1}^{N-1} \bar{c}_i \, ds_i = 0.$$

Because $\dim_{\mathbb{C}} V^* = N-1$ and t is a non-singular point on V^*, it follows from (4) that $s_1, \cdots, s_{N-1}$ are local parameters of V^* at t. But this contradicts (7). Q.E.D.

Corollary 1. If $\dim_{\mathbb{C}} V^* = N-1$, then f_2 is of degree one.

Proof. We use the same notations as in the proof of the Duality Theorem. We can assume that $S(V^*) \supseteq f_2(\tilde{V}-\Gamma')$ and $V^*-S(V^*)$ is non-singular. Let $t \in V^*-S(V^*)$, $z_1, z_2 \in f_2^{-1}(t)$, $x_i = f_1(z_i)$, $i = 1,2$. We have that $x_1, x_2 \in V_{sm}$ and $H^*_{x_1}, H^*_{x_2}$ are tangent to V^* at t. Since $\dim_{\mathbb{C}} V^* = N-1$ then $H^*_{x_1} = H^*_{x_2}$ and $x_1 = x_2$, $z_1 = z_2$. Q.E.D.

Corollary 2. If $\dim_{\mathbb{C}} V > 0$, $\dim_{\mathbb{C}} V^* = N-1$ and V has only isolated singular points then there exists a proper subvariety $S'(V^*) \supseteq S(V^*)$ such that for any $t \in V^*-S'(V^*)$ the corresponding hyperplane section E_t of V has only one singular point which is an ordinary quadratic singularity.

Proof. Extend $S(V^*)$ as in the proof of Corollary 1. Let $a_1, \cdots, a_N$ be all the singular points of V and $S'(V^*) = S(V^*) \cup (\bigcup_{i=1}^{N} (H^*_{a_i} \cap V^*))$. $\forall$ $i = 1,2,\cdots,N$ $H^*_{a_i} \not\supseteq V^*$ because $\dim_{\mathbb{C}} V^* = N-1$ and $H^*_{a_i} \supseteq V^*$ would mean that $V^* = H^*_{a_i}$ and $\dim_{\mathbb{C}} V = 0$ (V is dual to V^*). Thus $S'(V^*)$ is a proper subvariety of V^*. Let $t \in V^*-S'(V^*)$, x be a singular point of E_t. Because $x \in V_{sm}$ we have that $(x,t) \in \widetilde{V}$ (and $H_t \supset \tau_x$). As in Corollary 1 we see that x is unique on E_t. From Duality Theorem (4) we get that x is an ordinary quadratic singularity on E_t. Q.E.D.

Corollary 3. Let $\dim_{\mathbb{C}} V = 2$. Then $\dim_{\mathbb{C}} V < N-1$ iff there exists on V a 1-dimensional algebraic system of projective lines $\{\ell_\mu, \mu \subset M\}$ such that for any fixed generic $\mu \subset M$ and for all $x \in \ell_\mu \cap V_{sm}$ the corresponding τ_x does not depend on x and V is either a cone in $\mathbb{C}P^N$ or an algebraic surface, with singular locus of dimension one.

Proof. "If" part is evident. Consider the "only if" part.

It is evident that for $x \in V_{sm}$ $f_2 f_1^{-1}(x)$ is an N-3-dimensional projective subspace of $\mathbb{C}P^{N*}$. Denote $f_2 f_1^{-1}(x) = \mathbb{C}P^{N-3}(x)$. We have a rational map $g: V \longrightarrow G_N^{N-3}$ where G_N^{N-3} is the Grassmanian of all N-3-dimensional projective subspaces of $\mathbb{C}P^{N*}$.

Suppose $\dim_{\mathbb{C}} g(V) = 2$. That means that we have a 2-dimensional algebraic system $\{\mathbb{C}P^{N-3}(x), x \in V_{sm}\}$ on V^*. If $N = 3$ we would have $\dim V^* = 2 = N-1$. Thus $N > 3$. Because $\dim_{\mathbb{C}} V^* < N-1$ and $\bigcup_{x \in V_{sm}} \mathbb{C}P^{N-3}(x)$ is dense in V^* we have that for generic $t \in V^*$ there exists an algebraic curve D_t in V_{sm} such that $\forall x \in D_t$, $\mathbb{C}P^{N-3}(x) \ni t$. Let $\mathbb{C}P_t^4$ be a generic 4-dimensional projective subspace of $\mathbb{C}P^{N*}$ containing t, $V' = V^* \cap \mathbb{C}P^4_t$, $\lambda_x = \mathbb{C}P^4_t \cap \mathbb{C}P^{N-3}(x)$, $x \in D_t$. Let L be a maximal projective subspace of $\mathbb{C}P^{N*}$ contained in all $\mathbb{C}P^{N-3}(x)$, $x \in D_t$. It is clear that $\dim_{\mathbb{C}} L \leq N-4$. We can suppose that $\mathbb{C}P^4_t \cap L = t$. It follows from this that $\{\lambda_x\}$ is a 1-dimensional algebraic system of projective lines on V' passing through t. We can assume that t is non-singular on V* and (because of Bertini's Theorem) V' is irreducible and t is non-singular on V'. We get that V' is a non-singular cone and thus V' is a 2-plane in $\mathbb{C}P^{N*}$. We see that V* is an N-2-dimensional projective subspace of $\mathbb{C}P^{N*}$. Because V is dual to V*, V is a projective line in $\mathbb{C}P^N$. Contradiction.

Thus $\dim_{\mathbb{C}} g(V) < 2$. If $\dim_{\mathbb{C}} g(V) = 0$ then $V^* = \mathbb{C}P^{N-3}(x)$ and V is a 2-plane in $\mathbb{C}P^N$. We see that only the case which we have to consider now is $\dim_{\mathbb{C}} g(V) = 1$. There exist non-empty Zariski open subsets U of V_{sm} and $U_1 \subset g(V)$ such that g is defined in all points of U, $U_1 = g(U)$ and $\{g^{-1}(z) \cap U, z \in U_1\}$ is a 1-dimensional algebraic system of algebraic curves on U. Let $C_z = g^{-1}(z)$, $z \in U_1$. We have that $\mathbb{C}P^{N-3}(x)$ is the same for all $x \in C_z$. That means that τ_x is the same for all $x \in C_z$, and $\tau_x \supset C_z$, $x \in C_z$.

Because $\dim_{\mathbb{C}} V^* < N-1$ and V is dual to V* (Duality Theorem (3)) there exists a 1-dimensional algebraic system of projective lines $\{\ell_\mu, \mu \in M\}$ on V. Take generic $x \in U$ and let $\ell_{\mu(x)}, C_{z(x)}$

be some elements of $\{\ell_\mu, \mu \in M\}$ and $\{C_z, z \in U_1\}$ passing through x. Suppose that $\ell_{\mu(x)} \not\subset C_{z(x)}$. Let $\pi: V \longrightarrow \bar{V}$ be a generic projection of V in $\mathbb{C}P^3$ such that π is an isomorphism in some neighborhood of x. Let $\bar{x} = \pi(x)$, $\bar{\ell}_{\mu(x)} = \pi(\ell_{\mu(x)})$, $\bar{C}_{z(x)} = \pi(C_{z(x)})$, $\tau_{\bar{x}}$ be the tangent plane of $\bar{V}$ at $\bar{x}$. We have $\bar{\ell}_{\mu(x)} \cup \bar{C}_{z(x)} \subset \tau_{\bar{x}} \cap \bar{V}$. Because $\tau_{x'}$ is the same $\forall x' \in C_z$ we have that $\tau_{\bar{x}}$ is the same $\forall \bar{x}'$ in some neighborhood of $\bar{x}$ on $\bar{C}_{z(x)}$. That means that the hyperplane section divisor of $\bar{V}$ corresponding to $\tau_{\bar{x}}$ contains $\bar{C}_{z(x)}$ with the multiplicity greater than one. Hence the order of tangency of $\tau_{\bar{x}}$ and $\bar{V}$ at x is greater than one. Because $\bar{x}$ is generic on $\bar{V}$ we have that $\bar{V}$ is a 2-plane in $\mathbb{C}P^3$. Thus V is a 2-plane in $\mathbb{C}P^N$ and $\dim_{\mathbb{C}} g(V) = 0$. Contradiction.

We see that $\ell_{\mu(x)} \subset C_{z(x)}$, that is, for all $x' \in \ell_{\mu(x)} \cap V_{sm}$ the corresponding $\tau_{x'}$ is the same.

Now suppose that V has only isolated singular points and V is not a cone. Take generic line ℓ^* in $\mathbb{C}P^{N^*}$. Because $\dim_{\mathbb{C}} V^* < N-1$ we have $\ell^* \cap V^* = \emptyset$. We can assume that N-2-dimensional projective subspace $\mathbb{C}P^{N-2}(\ell^*)$ of $\mathbb{C}P^N$ dual to ℓ^* does not contain singular points of V. Let $B = \mathbb{C}P^{N-2}(\ell^*) \cap V$ and $x_o \in B$. Because V is not a cone we can assume that an element $\ell_\mu(x_o)$ of $\{\ell_\mu, \mu \in M\}$ passing through x_o does not contain singular points of V, that is, $\ell_{\mu(x_o)} \subset V_{sm}$. Take $x_1 \in \ell_{\mu(x_o)}$, $x_1 \neq x_o$. There exists a hyperplane H such that $H \supset \mathbb{C}P^{N-2}(\ell^*)$ and $H \supset x_1$. Then $H \supset \ell_\mu$. Let E be the hyperplane section of V corresponding to H. Because $E \supset \ell_\mu$, E is connected and evidently $E \neq \ell_\mu$ (otherwise, deg V = deg E = 1) we get that E has a singular point $x_2 \in \ell_\mu$. Since $x_2 \in V_{sm}$ we have that $H \supset \tau_{x_2}$. Thus the point $t(H) \in \mathbb{C}P^{N*}$ corresponding to H is contained in V* and $V^* \cap \ell^* \neq \emptyset$. Contradiction. Q.E.D.

<u>Corollary 4</u>. Let $\dim_{\mathbb{C}} V = 2$, $\dim_{\mathbb{C}} V^* = N-1$. Suppose that V is not contained in some 3-dimensional projective subspace of $\mathbb{C}P^N$. Then for generic $x,y \in V$, $\dim_{\mathbb{C}}(\tau_x \cap \tau_y) < 1$.

<u>Proof</u>. Let $\pi: V \longrightarrow \overline{V}$ be a generic projection of V in $\mathbb{C}P^4$. From Corollary 3 and $\dim_{\mathbb{C}} V^* = N-1$ we see that $\dim_{\mathbb{C}} \overline{V}^* = 3$. This shows that we can assume that $N = 4$.

Suppose that for generic $x,y \in V$ $\dim_{\mathbb{C}}(\tau_x \cap \tau_y) = 1$. That means that there exists a 3-dimensional projective subspace of $\mathbb{C}P^4$ containing τ_x and τ_y. Let $\mathbb{C}P^1(x) = f_2(f_1^{-1}(x))$, $x \in V_{sm}$. We see that for generic $x,y \in V$, $\mathbb{C}P^1(x) \cap \mathbb{C}P^1(y) \neq \emptyset$. Take generic $t \in V^*$. There exists a generic $x \in V$ with $\mathbb{C}P^1(x) \supset t$. For all generic $y \in V$ we have that $\mathbb{C}P^1(y) \cap \mathbb{C}P^1_{(x)} \neq \emptyset$. If the union of $\mathbb{C}P^1(y) \cap \mathbb{C}P^1(x)$ for fixed x and generic y would be a finite number of points on $\mathbb{C}P^1(x)$ we would have that there exists a point $b \in V^*$ such that $b \in \mathbb{C}P^1(y)$ for all generic $y \in V$. Let

H_b be the hyperplane of $\mathbb{C}P^4$ corresponding to b. We have $\tau_y \subset H_b$ for all generic $y \in V$. Thus $V \subset H_b$. Contradiction.

We get that there exists a non-empty Zariski open subset $U \subset \mathbb{C}P^1(x)$ such that $\forall a \in U$, $a \in \mathbb{C}P^1(x) \cap \mathbb{C}P^1(y)$ for some $y \in V_{sm}$. Considering the map $f_2: \tilde{V} \longrightarrow V^*$ and using $\dim_{\mathbb{C}} V^* = 3$ and Corollary 1 we see that all $a \in U$ are singular points of V^*. But taking t non-singular, we would have that $\mathbb{C}P^1(x)(\ni t)$ intersects with the singular part of V^* in only a finite number of points. Contradiction. Q.E.D.

<u>Corollary 5</u>. Let $\mathfrak{J}(V)$ be the algebraic closure in $\mathbb{C}P^N$ of the union of τ_x for all $x \in V_{sm}$. Suppose $\dim_{\mathbb{C}} V = 2$ and V is not in some 3-dimensional projective subspace of $\mathbb{C}P^N$.

Then $\dim_{\mathbb{C}} \mathfrak{J}(V) < 4$ if and only if $\dim_{\mathbb{C}} V^* < N-1$.

<u>Proof</u>. The "if" part immediately follows from Corollary 3. Consider the "only if" part. Using generic projection in $\mathbb{C}P^4$ and Corollary 3 we see that without loss of generality we can assume $N = 4$. Suppose $\dim_{\mathbb{C}} \mathfrak{J}(V) = 3$. $\{\tau_x, x \in V_{sm}\}$ is an algebraic system of 2-planes in $\mathfrak{J}(V)$. Let G be the Grassmanian of all 2-planes of $\mathbb{C}P^4$, $\pi: \mathcal{E} \longrightarrow G$ be the canonical $\mathbb{C}P^2$-bundle over G and $g: \mathcal{E} \longrightarrow \mathbb{C}P^4$ be the canonical map of $\mathcal{E}$ in $\mathbb{C}P^4$. We have a regular map $f: V_{sm} \longrightarrow G$. Let $\overline{\pi}: \overline{\mathcal{E}} \longrightarrow V_{sm}$ be the $\mathbb{C}P^2$-bundle over V_{sm} induced by $\pi: \mathcal{E} \longrightarrow G$ under f, $\tilde{f}: \overline{\mathcal{E}} \longrightarrow \mathcal{E}$ be the induced map, $T' = f(V_{sm})$ and T be the algebraic closure of T' in G. It is clear

that $\mathfrak{J}(V)$ is the algebraic closure of $g\tilde{f}(\bar{\mathfrak{C}})$ in $\mathbb{C}P^4$ and $\mathfrak{J}(V) = g(\pi^{-1}(T))$. We have an irreducible algebraic system $\{\mathbb{C}P^2(t), t \in T\}$ on $\mathfrak{J}(V)$ such that for generic $t \in T$, $\mathbb{C}P^2(t)$ is the tangent plane of V in some generic $x \in V$.

Consider two different possible cases.

<u>Case 1</u>. $\dim_{\mathbb{C}} T = 1$. Let $C_t = f^{-1}(t)$ for $t \in T'$. We have a 1-dimensional algebraic system of positive divisors on V. Let $\bar{C}_t = \text{Supp } C_t$, $\tilde{C}_t = \bar{C}_t \cap V_{sm}$. For generic $x \in V$ there exists a $\tilde{C}_t(x) \ni x$. Because $f(\tilde{C}_t)$ is a point in G we have that for all $y \in \tilde{C}_t(x)$ the corresponding tangent plane is the same. But that means that $\dim_{\mathbb{C}} V^* < N-1$.

<u>Case 2</u>. $\dim_{\mathbb{C}} T = 2$. In this case we have that for a generic $z \in \mathfrak{J}(V)$ there exist infinitely many different $\mathbb{C}P^2(t)$ passing through z. We can assume that for generic $t_1 \in T$ the corresponding $\mathbb{C}P^2(t_1)$ contains a non-singular point z of $\mathfrak{J}(V)$ with such property. Let $\mathbb{C}P^2(t_1) = \tau_x$ for some generic $x \in V_{sm}$ and $\mathbb{C}P^2(t_2)$ be another element of $\{\mathbb{C}P^2(t), t \in T\}$ passing through z. Then for all $t' \in T$ "close" to t_2 the corresponding $\mathbb{C}P^2(t')$ intersects $\mathbb{C}P^2(t_1)$ in some non-singular point z' on $\mathfrak{J}(V)$ "close" to z. We get that for generic $t \in T$, $\mathbb{C}P^2(t)$ intersects $\mathbb{C}P^2(t_1)$ in some point z(t) which is non-singular on $\mathfrak{J}(V)$. Let $y \in V_{sm}$ be such that $\mathbb{C}P^2(t) = \tau_y$. We obtain that for generic $x, y \in V$, $\tau_x \cap \tau_y$ contains a non-singular point of $\mathfrak{J}(V)$. Because

$\dim_{\mathbb{C}} J(V) = 3$ we have $\dim_{\mathbb{C}}(\tau_x \cap \tau_y) \geq 1$. From Corollary 4 we get $\dim_{\mathbb{C}} V^* < N-1$. Q.E.D.

Corollary 6. Let $\dim_{\mathbb{C}} V = 2$, $\dim_{\mathbb{C}} V^* = N-1$, suppose that V has only isolated singular points, V is not contained in a 3-dimensional projective subspace of $\mathbb{C}P^N$ and for generic $x \in V$ $\dim_{\mathbb{C}}[\tau_x \cap V] = 1$. Then for generic $x \in V$, $\tau_x \cap V$ is the union of some projective line of $\mathbb{C}P^N$ passing through x and of finite number of points.

Proof. It is easy to see that there exists an irreducible algebraic system of algebraic curves on V $\{C_t, t \in T\}$ and a rational map $f: V \longrightarrow T$ such that $f(V)$ is dense in T and for generic $x \in V$, $\tau_x(V)$ is the union of $C_{f(x)}$ and of finite number of points.

Consider two different possible cases.

Case 1. $\dim_{\mathbb{C}} T \leq 1$. Take generic $t \in T$ and let $D_t = f^{-1}(t) \cap V_{sm}$. Because $\dim_{\mathbb{C}} V^* = N-1$ we have that $\tau_{y_1} \neq \tau_{y_2}$ for generic $y_1, y_2 \in D_t$, $y_1 \neq y_2$. Because $C_t \subset \bigcup_{y \in D_t} \tau_y$ we see that C_t is a projective line. Take generic $x \in V$. There exists an element $C_{t(x)}$ of $\{C_t, t \in T\}$ passing through x. (If not, we would have a line $\ell \subset V$ such that all $\tau_x \supset \ell$, $x \in V_{sm}$. Then V^* is in some N-2-dimensional projective subspace of $\mathbb{C}P^{N*}$. This contradicts $\dim V^* = N-1$). Because $C_{t(x)}$ is a projective line we have $\tau_x \supset C_{t(x)}$ and $C_{t(x)} \subset \tau_x \cap V = C_{f(x)}$ + (finite number of points). Hence $C_{t(x)} = C_{f(x)}$ and $C_{f(x)} \ni x$.

<u>Case 2</u>. $\dim_{\mathbb{C}} T = 2$.

Suppose $\bigcap_{t \in T} C_t \neq \emptyset$ and let $a \in \bigcap_{t \in T} C_t$. Then for generic $x \in V$ we have $\tau_x \ni a$. Hence $V^* \subseteq H_a^*$ where H_a^* is the hyperplane of $\mathbb{C}P^{N*}$ corresponding to a. Because $\dim_{\mathbb{C}} V^* = N-1$ we have $V^* = H_a^*$. But V is dual to V^* and we get $\dim_{\mathbb{C}} V = 0$. Contradiction.

Thus $\bigcap_{t \in T} C_t = \emptyset$. Because V has only isolated singularities we see that for generic $t \in T$, $C_t \subset V_{sm}$. Corollary 2 gives us that for $z \in V^* - S'(V^*)$ the corresponding hyperplane section E_z has only one singular point $x(z)$ which is an ordinary quadratic singularity. It is clear that $C_{f(x(z))} \subset E_z$. If $C_{f(x(z))} = E_z$ we have that E_z is in a 2-plane of $\mathbb{C}P^N$. Hence V is in a 3-dimensional projective subspace of $\mathbb{C}P^N$. Contradiction.

Thus $C_{f(x(z))} \subsetneqq E_z$ and because E_z is connected we see that $C_{f(x(z))}$ is non-singular, $C_{f(x(z))} \ni x(z)$ and $(E_z - C_{f(x(z))})C_{f(x(z))} = 1$. In particular, we get that for generic $t \in T$, C_t is non-singular and irreducible. It easily follows from this and from $\dim_{\mathbb{C}} T = 2$

that $C_t^2 > 0$. Suppose $C_t^2 > 1$. Then we have two possibilities:
1) For generic $x,y \in V$, $x \neq y$, $C_{f(x)} \cap C_{f(y)}$ contains two different points. In that case we have $\dim_{\mathbb{C}}(\tau_x \cap \tau_y) \geq 1$ and Corollary 4 gives us a contradiction with $\dim_{\mathbb{C}} V^* = N-1$.
2) For generic $x,y \in V$, $x \neq y$, $C_{f(x)} \cap C_{f(y)}$ is one point $a(x,y)$ and $C_{f(x)}$ is tangent to $C_{f(y)}$ at $a(x,y)$. But that means that there is a projective line $\ell \subset \tau_{a(x,y)}$ which is the common tangent projective line in $\mathbb{C}P^N$ of $C_{f(x)}$ and $C_{f(y)}$ at $a(x,y)$. We get $\tau_x \supset \ell$, $\tau_y \supset \ell$ and $\dim_{\mathbb{C}}(\tau_x \cap \tau_y) \geq 1$. Again Corollary 4 gives us a contradiction with $\dim_{\mathbb{C}} V^* = N-1$. Thus $C_t^2 = 1$.

Take generic $x \in V_{sm}$ and consider the linear system $\{E_u\}$ of all hyperplane sections of V having singularity at x. Each $E_u \supset C_{f(x)}$ and let $D_u = E_u - C_{f(x)}$. Evidently each $D_u \ni x$. Take $y \in C_{f(x)} \cap V_{sm}$, $y \neq x$. Because $\{D_u\}$ is infinite (as $\{E_u\}(N > 3)$) there exists a $D_{u(y)} \ni y$. But $D_u . C_{f(x)} = 1$ and $x,y \in C_{f(x)} \cap D_{u(y)}$, $x \neq y$, give us that $C_{f(x)} \subseteq D_{u(y)}$ ($C_{f(x)}$ is irreducible). We can write $E_{u(y)} = 2C_{f(x)} + D'$ where D' is some non-negative divisor on V. Because Supp $E_{u(y)}$ is connected we have: if $D' \neq 0$ then $C_{f(x)} \cap D' \neq \emptyset$. Hence $E_{u(y)} . C_{f(x)} > 2$. But $E_{u(y)} . C_{f(x)} = D_u . C_{f(x)} + C_{f(x)} . C_{f(x)} = 2$. Contradiction.

We get $D' = 0$ and $E_{u(y)} = 2C_{f(x)}$. We showed that for any $t \in T$ there exists a hyperplane section $E(t)$ of V such that $E(t) = 2C_t$.

Because $\dim_{\mathbb{C}} T = 2$ we have infinite number of different C_t passing through x. Take one of them, say C_{t1}, $C_{t1} \neq C_{f(x)}$. Consider the corresponding $E(t_1) = 2C_{t_1}$. Then $E(t_1)$ is singular at x. Thus $E(t_1) \supset \tau_x \cap V$. We get $C_{t_1} \supset C_{f(x)}$. Contradiction.

Q.E.D.

§3. Proof of Theorem 3 (§3, Part I).

This theorem was formulated on page 60.

Proof of Theorem 3. Using generic projection $V \longrightarrow \mathbb{C}P^5$ we can assume that $N \leq 5$.

(1) Let K_{a_i}, $i = 1,2,\cdots,q$, be the algebraic closure in $\mathbb{C}P^N$ of the union of all projective lines $\ell(a_i,b)$ in $\mathbb{C}P^N$ connecting a_i and a point $b \in V-a_i$. Denote by M the parameter space of all projections $V \longrightarrow \mathbb{C}P^3$ (M is irreducible). It is clear that $\dim_{\mathbb{C}} K_{a_i} \leq 3$ and we can find a proper algebraic subvariety S_o in M such that for any $\pi \in M-S_o$ we have that π is regular on V, π is locally biregular at each a_i, $i = 1,2,\cdots,q$, and $\forall i = 1,2,\cdots,q$, $\pi^{-1}(\pi(a_i)) = a_i$. This finishes (1).

Let $\tilde{S}_o$ be a small open (classical) neighborhood of S_o in M. Because $M-\tilde{S}_o$ is compact, we can find open neighborhoods $\tilde{U}_i$ of a_i in V such that $\pi \in M-\tilde{S}_o$ and $\forall i = 1,2,\cdots,q$ we have that $\pi^{-1}(\pi(\tilde{U}_i)) = \tilde{U}_i$ and $\pi|_{\tilde{U}_i} : \tilde{U}_i \longrightarrow \pi(\tilde{U}_i)$ is a biregular map. Let $\tilde{V} = V - \bigcup_{i=1}^{q} \tilde{U}_i$.

(2) We can assume $N = 5$. (In the case $N = 4$ almost the same arguments work.) A projection $\pi \in M$ is defined by its center which is a projective line ℓ_π in $\mathbb{C}P^5$.

Denote by $\mathfrak{J}(V)$ the algebraic closure of the union of all tangent planes τ_x of V where x is a non-singular point of V. Let $T(V_{sm})$ be the tangent bundle of V_{sm} ($V_{sm} = V - \bigcup_{i=1}^{q} a_i$) and

$\psi: T(V_{sm}) \longrightarrow \mathcal{J}(V)$ be the canonical rational map. Define two integers a,b as follows:

If $\dim_{\mathbb{C}}\mathcal{J}(V) < 4$, then $a = 0$, $b = 0$, and if $\dim_{\mathbb{C}}\mathcal{J}(V) = 4$, then $a = \deg \mathcal{J}(V)$ (in $\mathbb{C}P^5$), $b = \deg \psi$.

Let $M_o = \{\pi \in M-\tilde{S}_o,\ \ell_\pi \cap V = \emptyset,\ \ell_\pi$ intersects $\mathcal{J}(V)$ in a different points and if $a \neq 0$, then for any $z \in \ell_\pi \cap \mathcal{J}(V)$ there exist b different points $x_i(z) \in V_{sm}$, $i = 1,2,\cdots,b$, such that $\forall i$ $z \in \tau_{x_i(z)}\}$. Clearly M_o is open and dense in $M-\tilde{S}_o$. Take $\pi_o \in M_o$. Let $\{z_1,\cdots,z_a\} = \ell_{\pi_o} \cap \mathcal{J}(V)$ and for any z_j, $j = 1,\cdots,a$ let x_{ij}, $i = 1,\cdots,b$, be all the points of V_{sm} with $\tau_{x_{ij}} \ni z_j$. Choose a hyperplane H_o in $\mathbb{C}P^5$ with $x_{ij} \notin H_o, i=1,\cdots,b,\ j=1,\cdots,a$, and $\overline{M}_o = \{\pi \in M_o,\ \ell_\pi \in H_o\}$. We shall prove the following

<u>Statement I</u>. Let x_o be an arbitrary element of $\{x_{ij},\ i = 1,2,\cdots,b,\ j = 1,\cdots,a\}$. There exists an open neighborhood U_{x_o} of x_o in V_{sm}, a neighborhood M_{x_o} of π_o in $\overline{M}_o$ and an open dense $M'_{x_o} \subset M_{x_o}$ such that for any $\pi' \in M'_{x_o}$ we have: for any $y \in U_{x_o}$ either $(d\pi')_y$ is a monomorphism or π' is a map of pinch-type at y (and last possibility holds only for a finite number of points of U_{x_o}).

<u>Proof of Statement I</u>. Let $\mathbb{C}^5_o = \mathbb{C}P^5 - H_o$. We can assume that affine coordinates $z_1,\cdots,z_5$ of $\mathbb{C}^5_o$ are chosen such that π_o is given by $(z_1,\cdots,z_5) \longrightarrow (z_1,z_2,z_3)$, $z_i(x_o) = 0$, $i = 1,2,\cdots,5$,

the hyperplane $z_3 = 0$ does not contain $T_{x_o} \cap \mathbb{C}_o^5$ and projection $(z_1,\cdots,z_5) \longrightarrow (z_3,z_4)$ is locally biregular on $V \cap \mathbb{C}_o^5$ at x_o. Let $\xi_i = z_i|_{V\cap\mathbb{C}_o^5}$, $i = 1,\cdots,5$. We see that ξ_3,ξ_4 are local parameters at x_o. There exists a Zariski open neighborhood U of x_o in $V \cap \mathbb{C}_o^5$ such that for any $y \in U$, $\xi_3-\xi_3(y)$, $\xi_4-\xi_4(y)$ are local parameters at y. Consider $\mathbb{C}^6$ with coordinates $(\epsilon_1,\epsilon_2,\epsilon_3,\epsilon_1',\epsilon_2',\epsilon_3')$ and let $z_i' = z_i+\epsilon_i z_4+\epsilon_i' z_5$, $i = 1,2,3$, $\xi_i' = z_i'|_{V\cap\mathbb{C}_o^5}$. For the Jacobians

$$\mathcal{J}_{13}' = \frac{D(\xi_1',\xi_3')}{D(\xi_3,\xi_4)} \quad \text{and} \quad \mathcal{J}_{23}' = \frac{D(\xi_2',\xi_3')}{D(\xi_3,\xi_4)}$$

we have:

$$\mathcal{J}_{13}' = \frac{D(\xi_1',\xi_3')}{D(\xi_3,\xi_4)} = \epsilon_1(-1-\epsilon_3'\frac{\partial\xi_5}{\partial\xi_3}) + \epsilon_3\frac{\partial\xi_1}{\partial\xi_3} + \epsilon_1'\epsilon_3\frac{\partial\xi_5}{\partial\xi_3} + \epsilon_3'\frac{D(\xi_1,\xi_5)}{D(\xi_3,\xi_4)} - \epsilon_1'\frac{\partial\xi_5}{\partial\xi_4} - \frac{\partial\xi_1}{\partial\xi_4},$$

$$\mathcal{J}_{23}' = \frac{D(\xi_2',\xi_3')}{D(\xi_3,\xi_4)} = \epsilon_2(-1-\epsilon_3'\frac{\partial\xi_5}{\partial\xi_3}) + \epsilon_3\frac{\partial\xi_2}{\partial\xi_3} + \epsilon_2'\epsilon_3\frac{\partial\xi_5}{\partial\xi_3} + \epsilon_3'\frac{D(\xi_2,\xi_5)}{D(\xi_3,\xi_4)} - \epsilon_2'\frac{\partial\xi_5}{\partial\xi_4} - \frac{\partial\xi_2}{\partial\xi_4}.$$

Consider $U \times \mathbb{C}^2$ and an algebraic variety W in $U \times \mathbb{C}^2$ defined by the equations:

$$\mathcal{J}_{13}'(\xi,\epsilon) = 0, \quad \mathcal{J}_{23}'(\xi,\epsilon) = 0$$

We can find open neighborhoods U_{x_o} of x_o in U and $\mathcal{E}$ of the origin in $\mathbb{C}^6$ such that in $U_{x_o} \times \mathcal{E}$

$$\frac{D(\xi_3', \xi_4)}{D(\xi_3, \xi_4)} \neq 0, \qquad \frac{D(\mathcal{J}_{13}', \mathcal{J}_{23}')}{D(\epsilon_1, \epsilon_2)} \neq 0,$$

that is, $W \cap [U_{x_o} \times \mathcal{E}]$ is non-singular and of complex dimension six. Let W_o be the irreducible component of W passing through the point $(x_o, 0)$, that is,

$$W_o \cap (U_{x_o} \times \mathcal{E}) = W \cap ((U_{x_o} \times \mathcal{E})) ,$$

$\alpha: W_o \longrightarrow \mathbb{C}^6$ be the map induced by projection, $\overline{\alpha(W_o)}$ be the algebraic closure of $\alpha(W_o)$ in $\mathbb{C}^6$. We have that either $\dim_{\mathbb{C}} \overline{\alpha(W_o)} < 6$ or there exists a proper algebraic subvariety S of $\mathbb{C}^6$ such that $\alpha|_{\alpha^{-1}(\mathbb{C}^6 - S)}: \alpha^{-1}(\mathbb{C}^6 - S) \longrightarrow \mathbb{C}^6 - S$ is an unramified map. In the case $\dim_{\mathbb{C}} \overline{\alpha(W_o)} < 6$ denote by $S = \overline{\alpha(W_o)}$. Let $\mathcal{E}' = \mathcal{E} \cap (\mathbb{C}^6 - S)$. We get that for any $(\epsilon_1, \epsilon_2, \epsilon_3, \epsilon_1', \epsilon_2', \epsilon_3') \in \mathcal{E}'$ the equations $\mathcal{J}_{13}' = 0$ and $\mathcal{J}_{23}' = 0$ define two complex subvarieties of U_{x_o}, say C_{13}', C_{23}', which in a neighborhood of any common point of them are non-singular complex curves intersecting transversally.

Fix any $(\epsilon_1, \epsilon_2, \epsilon_3, \epsilon_1', \epsilon_2', \epsilon_3') \in \mathcal{E}'$ and let $f: U_{x_o} \longrightarrow \mathbb{C}^3$ be the map defined by projection $(z_1, z_2, \ldots, z_5) \longrightarrow (z_1', z_2', z_3')$. Take any $y \in U_{x_o}$. If $y \notin C_{13}' \cap C_{23}'$ then $(df)_y$ is a monomorphism. Suppose $y \in C_{13}' \cap C_{23}'$. Let $u = \xi_3' - \xi_3'(y)$, $v = \xi_4 - \xi_4(y)$, $z_i = \xi_i' - \xi_i'(y)$, $i = 1,2$. We can write $z_i = \varphi_i(u,v)$, $i = 1,2$, where $\varphi_i(u,v)$ are some power series of u,v, $\varphi_i(0,0) = 0$. Using the fact $\frac{D(Z_i,u)}{D(u,v)}(y) = 0$, $i = 1,2$, and a linear transformation of type $Z_i \longrightarrow Z_i - a_i u$ we can assume that $\varphi_i(u,v)$ have no terms of degree one. Write $Z_i = \alpha_i u^2 + 2\beta_i uv + \gamma_i v^2$ + t.h.d. ("t.h.d." means "terms of higher degree".) We get from the transversality of C'_{12}, C'_{13} in y that $\det\begin{Vmatrix}\beta_1 & \gamma_1\\ \beta_2 & \gamma_2\end{Vmatrix} \neq 0$, and in particular one of γ_1, γ_2 is not zero. We can assume $\gamma_1 \neq 0$. Now writing $\varphi_1(u,v) = A(u,v)v^2 + B(u)v + C(u)$ and using $A(0,0) = \gamma_1 \neq 0$ we see that we can find new local parameters $u,\bar{v}$ such that $Z_1 = \bar{v}^2 + d(u)$, $d(u)$ has no terms of degree one, $d(0) = 0$, and $Z_2 = \bar{\alpha}u^2 + 2\bar{\beta}u\bar{v} + \bar{\gamma}\bar{v}^2$ + t.h.d., $\bar{\beta} \neq 0$. Let $\tilde{Z}_1 = Z_1 - d(u)$. We have now $\bar{v}^2 = \tilde{Z}_1$ and $Z_2 = \bar{A}(u,\tilde{Z}_1) + \bar{B}(u,\tilde{Z}_1)\bar{v}$, where $\bar{A}(u,\tilde{Z}_1)$ has no terms of degree one (and $\bar{A}(0,0) = 0$) and $\bar{B}(u,\tilde{Z}_1) = bu$ + t,h.d, $b \neq 0$. Let $\tilde{Z}_2 = Z_2 - \bar{A}(u,\tilde{Z}_1)$, $\tilde{Z}_3 = \bar{B}(u,\tilde{Z}_1)$, $\bar{u} = \bar{B}(u,\tilde{Z}_1)$. We see that $\bar{u},\bar{v}$ are local parameters at y and in some neighborhood of y the map f is given by the formulas:

$$\tilde{Z}_1 = \bar{v}^2 \ ;$$
$$\tilde{Z}_2 = \bar{u}\bar{v};$$
$$\tilde{Z}_3 = \bar{u},$$

that is, f is of pinch-type at y. Statement I is proved.

Let $\overline{M}_I = \bigcap_{\substack{i=1,\cdots,b \\ j=1,\cdots,a}} M_{x_{i,j}}$, $\overline{M}'_I = \bigcap_{\substack{i=1,\cdots,b \\ j=1,\cdots,a}} M'_{x_{i,j}}$. Now it is easy to see that for any point π of $\overline{M}'_I$ sufficiently close to π_o we have the following: for any $y \in V_{sm}$ either $(d\pi)_y$ is a monomorphism at y or π is a map of pinch-type at y. Choose such a π in $\overline{M}'_I$ and denote it by π_I. Because $\tilde{V}$ is compact we can choose an open neighborhood M_I of π in M_o such that for any $\pi \in M_I$ we have: for any $y \in V_{sm}$ either $(d\pi)_y$ is a monomorphism at y or π is a map of pinch-type at y.

Let $\dim_{\mathbb{C}} \mathfrak{J}(V) = 4$ (that is, $\dim V^* = N-1$ (Corollary 5 of the Duality Theorem)). For any $x \in V_{sm}$ let $K_{\tau,x}$ be the union of all projective lines of $\mathbb{C}P^5$ which connect x with other points of $\tau_x \cap V$ and $K_\tau(V)$ be the algebraic closure in $\mathbb{C}P^5$ of $\bigcup_{x \in V_{sm}} K_{\tau,x}$. From Corollary 6 of the Duality Theorem we get that for generic $x \in V$, $\tau_x \cap V$ is either a finite number of points or the union of a projective line in τ_x passing through x and of finite number of points. In both cases we have for generic $x \in V$, $\dim_{\mathbb{C}} K_{\tau,x} \leq 1$ and thus $\dim_{\mathbb{C}} K_\tau(V) \leq 3$. We can find $\pi' \in M_I$ such that $\ell_{\pi'} \cap K_\tau(V) = \emptyset$. Let $\ell_{\pi'} \cap \mathfrak{J}(X) = \{z'_1,\cdots,z'_a\}$, $z'_o = \emptyset$, and $x'_{1,j},\cdots,x'_{b,j}$ be all the points of V_{sm} with $\tau_{x'_{i,j}} \ni z'_j$ $(x_{i,o} = \emptyset)$ Suppose that for z'_j, $0 \leq j \leq k < a$ we already have the following property: $\forall\ i = 1,2,\cdots,b$, $\pi'^{-1}(\pi'(x'_{i,j})) = x'_{i,j}$.

For any $x \in V$ let K_x be the algebraic closure in $\mathbb{C}P^5$ of all projective lines connecting x with other points of V. Clearly $\dim_{\mathbb{C}} K_x \leq 3$ and for $x \in V_{sm}$, $\tau_x \subseteq K_x$. Let $K_{k+1} = \bigcup_{i=1}^{b} K_{x'_{i,K+1}}$.

We have $z'_{k+1} \in \ell_{\pi'} \cap K_{k+1}$. Because $\dim_{\mathbb{C}} K_{k+1} \leq 3$ we can find a π'' in M_I arbitrarily close to π' such that $\ell_{\pi''} \cap K_{k+1} = z'_{k+1}$. It is clear that $z'_{k+1} \in \ell_{\pi''} \cap \mathfrak{J}(V)$. Suppose for some $i = 1,2,\cdots,b$ there exists a point $y \in \pi''^{-1}\pi''(x'_{i,k+1})$ with $y \neq x'_{i,k+1}$. Let $P(\ell_{\pi''},x'_{i,k+1})$ be the 2-plane in $\mathbb{C}P^5$ containing $\ell_{\pi''}$ and $x'_{i,k+1}$ and $\ell(x'_{i,k+1},y)$ be the projective line containing $x_{i,k+1}$ and y. Clearly $\ell(x'_{i,k+1},y) \subset P(\ell_{\pi''},x'_{i,k+1})$ and there exists a point $z = \ell_{\pi''} \cap \ell(x'_{i,k+1},y)$. Because $\ell(x'_{i,k+1},y) \subset K_{x'_{i,k+1}}$ and $\ell_{\pi''} \cap K_{x'_{i,k+1}} = z'_{k+1}$ we have $z = z'_{k+1}$. Hence $z'_{k+1} \in \ell(x'_{i,k+1},y)$ and $\ell(x'_{i,k+1},y) \subset \tau_{x'_{i,k+1}}$. Thus $\ell(x'_{i,k+1},y) \subset K_{\tau,x'_{i,k+1}} \subset K_\tau(V)$ and $z'_{k+1} \in K_\tau(V)$. But this contradicts $\ell_{\pi''} \cap K_\tau(V) = \emptyset$. We see that $\pi''^{-1}(\pi''(x'_{i,k+1}) = x'_{i,k+1}$, $i = 1,2,\cdots,b$. Taking π'' sufficiently close to A' we can uniquely define the points $z''_j \in \ell_{\pi''} \cap \mathfrak{J}(V)$, $0 \leq j \leq K$, $x''_{i,j} \in V_{sm}$, $i = 1,2,\cdots,b$, $\tau_{x''_{i,j}} \ni z''_j$, such that z''_j are close to z'_j and $x''_{i,j}$ are close to $x'_{i,j}$. Because π'' is of pinch-type at all $x''_{i,j}$ we see that for π'' sufficiently close to π' we will have $\pi''^{-1}(\pi''(x''_{i,j})) = x''_{i,j}$, $0 \leq j \leq K$, $1 \leq i \leq b$.

This argument shows that we can choose such π_{II} in M_I that for any $y \in V_{sm}$ with the property: π_{II} is of pinch-type at y, we have $\pi_{II}^{-1}(\pi_{II}(y)) = y$. There exists an open neighborhood M_{II} of π_{II} in M_I such that for any $\pi \in M_{II}$ we have: If $y \in V_{sm}$ is such that π is of pinch-type at y then $\pi^{-1}(\pi(y)) = y$. Let E be a non-singular hyperplane section of V. For generic projection p: $E \longrightarrow \mathbb{C}P^2$ we know that p(E) has only ordinary double points. It follows from this that we can find a π_{III} in M_{II} such that there are only a finite number of points in $\pi_{III}(V)$ which are not ordinary singulariti We can assume also that for any $x \in \pi_{III}(V)$, $\pi_{III}^{-1}(x)$ is a finite set. There exists a neighborhood M_{III} of π_{III} in M_{II} such that all $\pi \in M_{III}$ have the same property as π_{III}.

Let H_1 be a hyperplane in $\mathbb{C}P^5$ which does not contain all such points x of V_{sm} that $\pi_{III}(x)$ is not an ordinary singularity of $\pi_{III}(V_{sm})$, $\bar{M}_{III} = \{\pi \in M_{III}, \ell_\pi \subset H_1\}$.

It is easy to verify that we will finish the proof of part (2) of our Theorem if we prove the following Statements and Corollaries.

Statement II. Let $x_1, x_2 \in V_{sm}$, $x_1 \neq x_2$, $\pi_{III}(x_1) = \pi_{III}(x_2)$ and $\pi_{III}(\tau_{x_1}) = \pi_{III}(\tau_{x_2})$. Then there exist open neighborhoods U_{x_i} of x_i, i = 1,2, in V_{sm}, a neighborhood M_{x_1,x_2} of π_{III} in $\bar{M}_{III}$ and an open dense $M'_{x_1,x_2} \subset M_{x_1,x_2}$ such that for any $\pi' \in M'_{x_1,x_2}$ we have:

if $y_1 \in U_{x_1}$, $y_2 \in U_{x_2}$ are such that $\pi'(y_1) = \pi'(y_2)$, $y_1 \neq y_2$, then $\pi'(\tau_{y_1}) \neq \pi'(\tau_{y_2})$.

Corollary of Statement II. There exists a $\pi_{IV} \in \overline{M}_{III}$ such that for any pair $y_1, y_2 \in V_{sm}$ with $\pi_{IV}(y_1) = \pi_{IV}(y_2)$, $y_1 \neq y_2$, we have: $\pi_{IV}(\tau_{y_1}) \neq \pi_{IV}(\tau_{y_2})$. Moreover, we can find an open neighborhood $\overline{M}_{IV}$ of π_{IV} in $\overline{M}_{III}$ such that for any $\pi \in \overline{M}_{IV}$ we have: if $y_1, y_2 \in V_{sm}$ are such that $\pi(y_1) = \pi(y_2)$, $y_1 \neq y_2$ then $\pi(\tau_{y_1}) \neq \pi(\tau_{y_2})$.

Statement III. Let x_1, x_2, x_3 be three different points of V_{sm} with $\pi_{IV}(x_1) = \pi_{IV}(x_2) = \pi_{IV}(x_3)$ and $\dim_{\mathbb{C}}\left[\bigcap_{i=1}^{3} \pi_{IV}(\tau_{x_i})\right] = 1$. Then there exist open neighborhoods U_{x_i} of x_i, $i = 1,2,3$, in V_{sm}, a neighborhood M_{x_1,x_2,x_3} of π_{IV} in $\overline{M}_{IV}$ and open dense $M'_{x_1,x_2,x_3} \subset M_{x_1,x_2,x_3}$ such that for any $\pi' \in M'_{x_1,x_2,x_3}$ we have:

if $y_i \in U_{x_i}$, $i = 1,2,3$, are such that $y_1 \neq y_2$, $y_2 \neq y_3$, $y_1 \neq y_3$ and $\pi'(y_1) = \pi'(y_2) = \pi'(y_3)$ then $\dim_{\mathbb{C}} \bigcap_{i=1}^{3} \pi'(\tau_{y_1}) = 0$.

Corollary of Statement III. There exists a $\pi_V \in \overline{M}_{IV}$ such that if $y_1, y_2, y_3 \in V_{sm}$, $\pi_V(y_1) = \pi_V(y_2) = \pi_V(y_3)$ and $y_1 \neq y_2$, $y_2 \neq y_3$, $y_1 \neq y_3$ then $\dim_{\mathbb{C}} \bigcap_{i=1}^{3} \pi_V(\tau_{y_2}) = 0$. Moreover we can find an open neighborhood $\overline{M}_V$ of π_V in $\overline{M}_{IV}$ such that for any $\pi \in \overline{M}_V$ we have:

if $y_1, y_2, y_3 \in V_{sm}$ are such that $\pi(y_1) = \pi(y_2) = \pi(y_3)$,
$y_1 \neq y_2$, $y_2 \neq y_3$; $y_1 \neq y_3$ then $\dim_{\mathbb{C}} \bigcap_{i=1}^{3} \pi(\tau_{y_i}) = 0$.

<u>Statement IV</u>. Let x_1, x_2, x_3, x_4 be four different points of V_{sm} with $\pi_V(x_1) = \pi_V(x_2) = \pi_V(x_3) = \pi_V(x_4)$.

Then there exist open neighborhoods U_{x_i} of x_i, $i = 1,2,3,4$, in V_{sm}, a neighborhood $M_{x_1,\ldots,x_4}$ of π_V in $\bar{M}_V$ and open dense

$$M'_{x_1,\ldots,x_4} \subset M_{x_1,\ldots,x_4}$$

such that for any $\pi' \in M'_{x_1,\ldots,x_4}$ we have $\bigcap_{i=1}^{4} \pi'(U_{x_i}) = \emptyset$.

<u>Corollary of Statement IV</u>. There exists a $\pi_{VI} \in \bar{M}_V$ such that for any $x \in \pi_{VI}(V)$ we have: $\pi_{VI}^{-1}(x)$ has less than four elements.

Note that all Corollaries immediately follow from the Statements (arguments are the same as used for the construction of π_I (page 104)).

<u>Proof of Statement II</u>.

We can choose affine coordinates in $\mathbb{C}_1^5 = \mathbb{C}P^5 - H_1$ such that π_{III} on $V \cap \mathbb{C}_1^5$ is given by $(z_1,\ldots,z_5) \longrightarrow (z_1,z_2,z_3)$, $z_i(\pi_{III}(x_1)) = z_i(\pi_{III}(x_2)) = 0$, $i = 1,2,3$, $z_1|_V$, $z_2|_V$ are local parameters of V at x_1 and x_2, $z_3 = 0$ on $\tau_{x_1} \cap \mathbb{C}_1^5$ and on $\tau_{x_2} \cap \mathbb{C}_1^5$. We can find open neighborhoods U_i of x_i in $\mathbb{C}_1^5$, $i = 1,2$, such that $V \cap U_i$ is given in U_i by equations:

$$z_3 = A_i(z_1,z_2),\ z_4 = B_i(z_1,z_2), z_5 = C_i(z_1,z_2),\ i = 1,2,$$

where A_i, B_i, C_i are power series of z_1, z_2, $A_i(0) = 0$, A_i has no terms of degree one, $i = 1,2,3$, and one of two numbers $B_1(0)-B_2(0)$, $C_1(0)-C_2(0)$ is not zero (we use $x_1 \neq x_2$). Consider $\mathbb{C}^6$ with coordinates $(\epsilon_1,\epsilon_2,\epsilon_3,\epsilon_1',\epsilon_2',\epsilon_3')$ and let $z_j' = z_j+\epsilon_j z_4+\epsilon_j' z_5$, $j = 1,2,3$. Consider $\mathbb{C}^2$ with coordinates z_1', z_2'.

Taking smaller U_1 and U_2 we can choose an open neighborhood U of (o) in $\mathbb{C}^2 \times \mathbb{C}^6$ such that there exist holomorphic functions $\tilde{A}_i$ in U with the following property: $z_3' = \tilde{A}_i(z_1', z_2'; \epsilon_1, \ldots, \epsilon_3')$ on $V \cap U_i$, $i = 1,2$. Let W be a complex-analytic subvariety in U given by the equation:

$$F = \tilde{A}_1(z_1', z_2', \epsilon_1, \ldots, \epsilon_3') - \tilde{A}_2(z_1', z_2', \epsilon_1, \ldots, \epsilon_3') = 0.$$

It is easy to verify that $\frac{\partial F}{\partial \epsilon_3}(0) = B_1(0)-B_2(0)$, $\frac{\partial F}{\partial \epsilon_3'}(0) = C_1(0)-C_2(0)$.

It follows from this that taking U smaller we can assume that W is 7-dimensional and non-singular. Let $\alpha\colon W \longrightarrow \mathbb{C}^6$ be the map induced by projection. Considering $\mathbb{C}^5_1 \times \mathbb{C}^5_1 \times \mathbb{C}^6 \longrightarrow \mathbb{C}^6$ it is not difficult to construct an algebraic variety $\overline{W}$ and a regular map $\overline{\alpha}\colon \overline{W} \longrightarrow \mathbb{C}^6$ such that $W \subset \overline{W}$, W is open in $\overline{W}$ and $\alpha = \overline{\alpha}|_W$. We see that there exists a neighborhood $\mathcal{E}$ of (o) in $\mathbb{C}^6$ and a proper analytic subvariety S in $\mathcal{E}$ such that either $\overline{\alpha}'(u) = \emptyset$ $\forall u \in \mathcal{E}-S$ or $\forall u \in \mathcal{E}-S$ and $\forall v \in \alpha^{-1}(u)$ $(d\alpha)_v$ is an epimorphism. This remark finishes the proof of Statement II.

Proof of Statement III.

We can choose affine coordinates in $\mathbb{C}_1^5$ such that π_{IV} on $V \cap \mathbb{C}_1^5$ is given by $(z_1,\cdots,z_5) \longrightarrow (z_1,z_2,z_3)$, $z_i(\pi_{IV}(x_1)) = z_i(\pi_{IV}(x_2)) = z_i(\pi_{IV}(x_3)) = 0$, $i = 1,2,3$, $z_4(x_3) = z_5(x_3) = 0$, $z_i = 0$ on $\tau_{x_i} \cap \mathbb{C}_1^5$, $i = 1,2$, $z_1+z_2 = 0$ on $\tau_{x_3} \cap \mathbb{C}_1^5$, $z_2|_V$, $z_3|_V$ (corresp. $z_1|_V$, $z_3|_V$) are local parameters of V at x_1 (corresp. at x_2 and x_3). We can find open neighborhoods U_i of x_i in $\mathbb{C}_1^5$, $i = 1,2,3$, such that $V \cap U_i$ is given in U_i by equations:

for $i = 1$: $z_1 = A_1(z_2,z_3)$, $z_4 = B_1(z_2,z_3)$, $z_5 = C_1(z_2,z_3)$,

for $i = 2$: $z_2 = A_2(z_1,z_3)$; $z_4 = B_2(z_1,z_3)$, $z_5 = C_2(z_1,z_3)$,

for $i = 3$: $z_1+z_2=A_3(z_1,z_3)$; $z_4 = B_3(z_1,z_3)$, $z_5 = C_3(z_1,z_3)$,

where A_i,B_i,C_i are power series of the corresponding variables, $A_i(0) = 0$, A_i has no terms of degree one, $i = 1,2,3$, $B_3(0) = C_3(0) = 0$ and one of four numbers $B_1(0)\cdot B_2(0)$, $B_1(0)\cdot C_2(0)$, $B_2(0)\cdot C_1(0)$, $C_1(0)\cdot C_2(0)$ is not equal to zero (we use that all x_1,x_2,x_3 are different).

Consider $\mathbb{C}^6$ with coordinates $(\epsilon_1,\epsilon_2,\epsilon_3,\epsilon_1',\epsilon_2',\epsilon_3')$ and let $z_j' = z_j+\epsilon_j z_4+\epsilon_j z_5$, $j = 1,2,3$.

Taking smaller U_i, $i = 1,2,3$, we can choose a positive number r such that there exist holomorphic functions

$\tilde{A}_1(z_2',z_3',\epsilon_1,\epsilon_2,\epsilon_3,\epsilon_1',\epsilon_2',\epsilon_3')$, $\tilde{A}_2(z_1',z_3',\epsilon_1,\epsilon_2,\cdots,\epsilon_3')$, $\tilde{A}_3(z_1',z_3',\epsilon_1,\cdots,\epsilon_3')$ defined for $|z_i'| < r$, $|\epsilon_i| < r$, $|\epsilon_i'| < r$, $i = 1,2,3$, which have the following property:

$$z_1' = \tilde{A}_1(z_2',z_3',\epsilon_1,\cdots,\epsilon_3') \quad \text{on} \quad V \cap U_1,$$

$$z_2' = \tilde{A}_2(z_1',z_3',\epsilon_1,\cdots,\epsilon_3') \quad \text{on} \quad V \cap U_2,$$

$$z_1'+z_2' = \tilde{A}_3(z_1',z_3',\epsilon_1,\cdots,\epsilon_3') \quad \text{on} \quad V \cap U_3.$$

Let $U = \{z_1',z_2',z_3',\epsilon_1,\cdots,\epsilon_3') \in \mathbb{C}^3 \times \mathbb{C}^6,\ |z_i'|<r,\ |\epsilon_i|<r,\ |\epsilon_i'|<r\}$ and W be a complex-analytic subvariety in U given by the equations:

$$F_1 = z_1' - \tilde{A}_1(z_2',z_3',\epsilon_1,\cdots,\epsilon_3') = 0$$

$$F_2 = z_2' - \tilde{A}_2(z_1',z_3',\epsilon_1,\cdots,\epsilon_3') = 0$$

$$F_3 = z_1' + z_2' - \tilde{A}_3(z_1',z_3',\epsilon_1,\cdots,\epsilon_3') = 0.$$

It is easy to verify that the matrix

$$\left\|\begin{matrix} \frac{\partial F_1}{\partial z_1'} & \frac{\partial F_1}{\partial \epsilon_1} & \frac{\partial F_1}{\partial \epsilon_2} & \frac{\partial F_1}{\partial \epsilon_1'} & \frac{\partial F_1}{\partial \epsilon_2'} \\ - & - & - & - & - \\ \frac{\partial F_3}{\partial z_1'} & - & - & - & \frac{\partial F_3}{\partial \epsilon_2'} \end{matrix}\right\|$$

has at the point (0) the following form

$$\left\|\begin{matrix} 1 & B_1(0) & 0 & C_1(0) & 0 \\ 0 & 0 & B_2(0) & 0 & C_2(0) \\ 1 & 0 & 0 & 0 & 0 \end{matrix}\right\|.$$

It follows from this that taking r smaller we can assume that W is 6-dimensional and non-singular. Now we have only to repeat almost word by word the last part of the proof of Statement II.

Proof of Statement IV.

We can choose affine coordinates in $\mathbb{C}_1^5$ such that π_V on $V \cap \mathbb{C}_1^5$ is given by $(z_1,\dots,z_5) \longrightarrow (z_1,z_2,z_3)$, $z_j(\pi_V(x_i)) = 0$, $j = 1,2,3$, $i = 1,2,3,4$, $z_4(x_4) = z_5(x_4) = 0$, $z_j = 0$ on $\tau_{x_j} \cap \mathbb{C}_1^5$, $j = 1,2,3$, $z_1+z_2+z_3 = 0$ on $\tau_{x_4} \cap \mathbb{C}_1^5$, $z_{j'}|_V, z_{j''}|_V$ (corresp. $z_1|_V, z_2|_V$) are local parameters of V at x_j, $j = 1,2,3$, $(j',j'') = (1,2,3)-(j)$, (corresp. x_4). We can find open neighborhoods U_j of x_j in $\mathbb{C}_1^5$, $j = 1,2,3,4$, such that $V \cap U_j$ is given in U by equations:

$$\text{for } j = 1,2,3\text{: } z_j = A_j(z_{j'},z_{j''}),\ z_4 = B_j(z_{j'},z_{j''}),\ z_5 = C_j(z_{j'},z_{j''})$$

$$\text{for } j = 4\text{: } z_1+z_2+z_3 = A_4(z_1,z_2),\ z_4 = B_4(z_1,z_2),\ z_5 = C_4(z_1,z_2)$$

where for $j = 1,2,3,4$ A_j, B_j, C_j are power series of the corresponding variables, $A_j(0) = 0$, A_j has no terms of degree one, $B_4(0) = C_4(0) = 0$ and the rank of the matrix

$$\left\|\begin{matrix} B_1(0) & 0 & 0 & C_1(0) & 0 & 0 \\ 0 & B_2(0) & 0 & 0 & C_2(0) & 0 \\ 0 & 0 & B_3(0) & 0 & 0 & C_3(0) \end{matrix}\right\|$$

is equal three (we use that all x_1,x_2,x_3,x_4 are different). Consider $\mathbb{C}^6$ with coordinates $(\epsilon_1,\dots,\epsilon_3')$ and let

$z_j' = z_j + \epsilon_j z_4 + \epsilon_j' z_5$, $j = 1,2,3$. Taking U_i, $i = 1,2,3$ smaller, we can choose a positive number r such that there exist holomorphic functions $\widetilde{A}_j(z_{j'}', z_{j''}', \epsilon_1, \cdots, \epsilon_3')$, $j = 1,2,3$, $(j',j'') = (1,2,3)-(j)$, and $\widetilde{A}_4(z_1', z_2', \epsilon_1, \cdots, \epsilon_3')$ defined for $|z_i'| < r$, $|\epsilon_i| < r$, $|\epsilon_i'| < r$, $i = 1,2,3$, which have the following property:

$$\text{for } j = 1,2,3 \quad z_j' = \widetilde{A}_j(z_{j'}', z_{j''}', \epsilon_1, \cdots, \epsilon_3') \text{ on } V \cap U_j$$

$$\text{and for } j = 4 \quad z_1' + z_2' + z_3' = \widetilde{A}_4(z_1', z_2', \epsilon_1, \cdots, \epsilon_3') \text{ on } V \cap U_4.$$

Let $U = \{(z_1', z_2', z_3', \epsilon_1, \cdots, \epsilon_3') \in \mathbb{C}^3 \times \mathbb{C}^6,\ |z_i'| < r,\ |\epsilon_i| < r,\ |\epsilon_i'| < r\}$

and W be a complex-analytic subvariety of U defined by the equations:

$$F_j = z_j' - \widetilde{A}_j(z_{j'}', z_{j''}', \epsilon_1, \cdots, \epsilon_3') = 0, \quad j = 1,2,3,$$

$$F_4 = z_1' + z_2' + z_3' - \widetilde{A}_4(z_1', z_2', \epsilon_1, \cdots, \epsilon_3') = 0.$$

It is easy to verify that the matrix

$$\left\| \begin{matrix} \dfrac{\partial F_1}{\partial z_1'} & \dfrac{\partial F_1}{\partial \epsilon_1} & \dfrac{\partial F_1}{\partial \epsilon_2} & \text{---------} & \dfrac{\partial F_1}{\partial \epsilon_3'} \\ \vdots & \vdots & \vdots & \text{---------} & \vdots \\ \dfrac{\partial F_4}{\partial z_1'} & \dfrac{\partial F_4}{\partial \epsilon_1} & & \text{-------------} & \dfrac{\partial F_4}{\partial \epsilon_3'} \end{matrix} \right\|$$

has at the point (0) the following form:

$$\left\| \begin{matrix} 1 & B_1(0) & 0 & 0 & C_1(0) & 0 & 0 \\ 0 & 0 & B_2(0) & 0 & 0 & C_2(0) & 0 \\ 0 & 0 & 0 & B_3(0) & 0 & 0 & C_3(0) \\ 1 & 0 & 0 & 0 & 0 & 0 & 0 \end{matrix} \right\|$$

and thus it has rank 4.

It follows from this that taking r smaller we can assume that W is 5-dimensional. Let $\alpha: W \longrightarrow \mathbb{C}^6$ be the map induced by projection. It is not difficult to construct an algebraic variety $\overline{W}$ and a regular map $\overline{\alpha}: \overline{W} \longrightarrow \mathbb{C}^6$ such that $W \subset \overline{W}$, W is open in $\overline{W}$ and $\alpha = \overline{\alpha}|_W$. We see that there exists a neighborhood $\mathcal{E}$ of (0) in $\mathbb{C}^6$ and a proper analytic subvariety S in $\mathcal{E}$ such that $\forall u \in \mathcal{E}-S$, $\alpha^{-1}(u) = \emptyset$. This remark finishes the proof of the Statement IV. We proved part (2) of the Theorem 1.

(3) (D. Mumford). It is easy to see that from the condition of the Theorem it follows that V is not a cone. From Severi's Theorem (page 72) we get that K(V) has (complex) dimension 5. Let G be the Grassmanian of projective lines in $\mathbb{C}P^5$ and

$$\begin{array}{ccc} \mathbb{C}P^5 & \xleftarrow{\;\pi_2\;} & \mathcal{E} \\ & & \downarrow{\scriptstyle \pi_1} \\ & & G \end{array}$$

be the canonical diagram corresponding to G ($\pi_1: \mathcal{E} \longrightarrow G$ is the canonical line bundle). Constructing chords we get a regular map $f: V \times V - \Delta \longrightarrow G$ (Δ is the diagonal). Let $\pi_1': \mathcal{E}' \longrightarrow V \times V - \Delta$ be the line bundle induced by π_1 under f and

$$\begin{array}{ccc} \mathcal{E}' & \xrightarrow{\ F\ } & \mathcal{E} \\ {\scriptstyle \pi_1'}\downarrow & & \downarrow{\scriptstyle \pi_1} \\ V\times V-\Delta & \xrightarrow[\ f\ ]{} & G \end{array}$$

be the induced diagram. Evidently $K(V)$ is the algebraic closure in $\mathbb{C}P^5$ of $\pi_2 F(\mathcal{E}')$. Because $\dim_{\mathbb{C}} K(V) = 5$, we have $K(V) = \mathbb{C}P^5$. Now considering $\pi_2 F\colon \mathcal{E}' \longrightarrow \mathbb{C}P^5$ and using Bertini's Theorem we get that for generic line $\ell \subset \mathbb{C}P^5$ $(\pi_2 F)^{-1}(\ell)$ is an irreducible algebraic curve in $\mathcal{E}'$. Let $p_1'\colon V\times V-\Delta \longrightarrow V$ be the map induced by projection on the first factor. It is easy to verify that

(*) $p_1'\pi_1'(\pi_2 F)^{-1}(\ell) = \{x \in V \mid \exists\, y \in V,\ y \neq x,\ \pi_\ell(y) = \pi_\ell(x)\}$,

where $\pi_\ell\colon V \longrightarrow \mathbb{C}P^3$ is the projection with center ℓ. Let $\tilde{C}$ be the algebraic closure of $p_1'\pi_1'(\pi_2 F)^{-1}(\ell)$ in V. We can assume that ℓ is such that π_ℓ satisfies to the parts (1) and (2) of the Theorem. Then we get from (*) that $\tilde{C} = \pi_\ell^{-1}(S_{\pi_\ell}(V))$. Thus $\pi_\ell^{-1}(S_{\pi_\ell}(V)$ and $S_{\pi_\ell}(V)$ are irreducible algebraic curves. We proved part (3) of Theorem 1.

(4) (D. Mumford) From Corollary 5 and 3 of the Duality Theorem we get that $\dim_{\mathbb{C}} \mathcal{J}(V) = 4$. Let $\pi\colon V \longrightarrow \mathbb{C}P^3$ be a projection which already satisfies the parts (1) and (2) of the Theorem and $Z_\pi \subset \mathbb{C}P^N$ be the center of π. Because $\operatorname{codim}_{\mathbb{C}} Z_\pi = 4$ we have that $Z_\pi \cap \mathcal{J}(V) \neq \emptyset$. But that means that $\pi(V)$ has pinch points. Q.E.D.

PART II

ELLIPTIC SURFACES

§1. Deformations of elliptic surfaces with "non-stable" singular fibers.

Definition. Let $f\colon V \longrightarrow \Delta$ be a proper holomorphic map of a (non-singular) complex surface V to a non-singular compact complex curve Δ such that for generic $x \in \Delta$, $f^{-1}(x)$ is a (non-singular) elliptic curve and there are no exceptional curves in the fibers of f. Following Kodaira we call $f\colon V \longrightarrow \Delta$ an analytic fiber space of elliptic curves or an elliptic fibration.

Theorem 8. Let $f\colon V \longrightarrow \Delta$ be an elliptic fibration which has singular multiple fibers ("singular multiple fiber" means "multiple fiber which (being reduced) is a singular curve"). Then there exists a commutative diagram of proper surjective holomorphic maps of complex manifolds

$$\begin{array}{ccc} & W & \\ {}^{h}\swarrow & & \searrow^{F} \\ \mathbb{D} & \xleftarrow{\;p\;} & W_1 \end{array}$$

where $\mathbb{D} = \{\tau \in \mathbb{C} \mid |\tau| \leq 1\}$, such that $F|_{h^{-1}(0)}\colon h^{-1}(0) \longrightarrow p^{-1}(0)$ coincides with $f\colon V \longrightarrow \Delta$, for any $\tau \in \mathbb{D}-(0)$, $F|_{h^{-1}(\tau)}\colon h^{-1}(\tau) \longrightarrow p^{-1}(\tau)$ is an elliptic fibration without singular multiple fibers and h has no critical values. Moreover, the number $r(\tau)$ of multiple fibers of $F|_{h^{-1}(\tau)}$ and the set of corresponding multiplicities $\{m_1(\tau), m_2(\tau), \cdots, m_{r(\tau)}(\tau)\}$ do not depend on τ $(\tau \neq 0)$.

Proof. The Theorem follows from the following

Lemma 4. Let $f'\colon V' \longrightarrow \mathcal{D}$ be a proper surjective map of complex manifolds with a single critical value $(0) \in \mathcal{D}$ such that for any $\sigma \in \mathcal{D}-(0)$, $f'^{-1}(\sigma)$ is a non-singular elliptic curve and $f'^{-1}(0)$ is a fiber of type ${}_mI_b$, $m \geq 2$, $b \geq 1$ (see [13]). Then

1) There exists $\epsilon, \epsilon' \in \mathbb{R}$, $\epsilon' > 0$, $0 < \epsilon < 1$, and a commutative diagram of surjective holomorphic maps of complex manifolds

(1)
$$\begin{array}{ccc} & W' & \\ {}^{h'}\swarrow & & \searrow^{F'} \\ \mathcal{D}_{\epsilon'} & \xleftarrow{\;\;p'\;\;} & \mathcal{D}_\epsilon \times \mathcal{D}_{\epsilon'} \end{array}$$

where $\mathcal{D}_{\epsilon'} = \{\tau \in \mathbb{C} \,|\, |\tau| < \epsilon'\}$, $\mathcal{D}_\epsilon = \{\sigma \in \mathbb{C} \,|\, |\sigma| < \epsilon\}$ $(\mathcal{D}_\epsilon \subset \mathcal{D} = \mathcal{D}_1)$, $p'\colon \mathcal{D}_\epsilon \times \mathcal{D}_{\epsilon'} \longrightarrow \mathcal{D}_{\epsilon'}$ is the projection, F' is a proper map such that $F'|_{h'^{-1}(0)}\colon h'^{-1}(0) \longrightarrow p'^{-1}(0)$ coincides with $f'|_{f'^{-1}(\mathcal{D}_\epsilon)}\colon f'^{-1}(\mathcal{D}_\epsilon) \longrightarrow \mathcal{D}_\epsilon$, for any $\tau \in \mathcal{D}_{\epsilon'}-0$, $F'|_{h'^{-1}(\tau)}\colon h'^{-1}(\tau) \longrightarrow p'^{-1}(\tau)$ has as generic fiber a non-singular elliptic curve, $F'^{-1}(0,\tau)$ is the unique multiple fiber of $F'|_{h'^{-1}(\tau)}\colon h'^{-1}(\tau) \longrightarrow p'^{-1}(\tau)$ which is of type ${}_mI_0$ (that is, non-singular) and h' has no critical values;

2) there exist $\epsilon_1, \epsilon_2 \in \mathbb{R}$, $0 < \epsilon_2 < \epsilon_1 < \epsilon$ and a commutative diagram of holomorphic maps

(2)

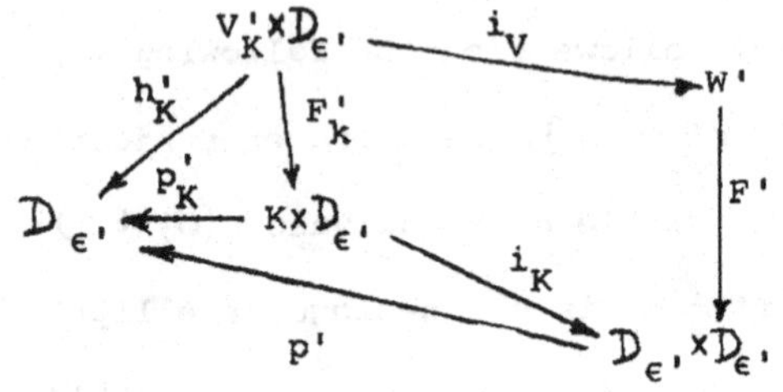

where $K = \{\sigma \in \mathbb{C} \mid \epsilon_2 < |\sigma| < \epsilon_1\}$, i_V and i_K are biholomorphic embeddings, h'_K and p'_K are evident projections, $V'_k = f'^{-1}(K)$,

$F'_K = (f'|_{V'_K}) \times (\text{identity})$, $i_K|_{K\times 0}: K\times 0 \longrightarrow D_\epsilon \times 0$

coincides with the natural embedding $K \longrightarrow D_\epsilon$ and

$i_V|_{V'_k \times 0}: V'_K \times 0 \longrightarrow F'^{-1}(D_\epsilon \times 0) (= V')$ coincides with the natural embedding $V'_k \longrightarrow V'$.

Suppose for a moment that Lemma 4 is already proved. Then the proof of Theorem 8 proceeds as follows:

Let $a_1, \cdots, a_r$ be all the points of Δ such that $\forall\, j = 1,2,\cdots$ $f^{-1}(a_i)$ is a singular multiple fiber, $D_{\epsilon,j}$ be a small open neighborhood of a_j in Δ such that $D_{\epsilon,j}$ is isomorphic to D_ϵ and $\forall\, t \in D_{\epsilon,j} - a_j$, $f^{-1}(t)$ are regular fibers of $f: V \longrightarrow \Delta$, $V'_j = f^{-1}(D_{\epsilon,j})$, $f'_j = f|_{V'_j}: V'_j \longrightarrow D_{\epsilon,j}$. Using Lemma 4 and takin

ϵ sufficiently small, we construct for all $f_j: V'_j \longrightarrow \mathcal{D}_{\epsilon,j}$ the diagrams analogous to the diagrams (1) and (2). We shall use the same notations as in Lemma 4 for the new objects which we obtain adding only index j (that is, we have $W'_j, F'_j, h'_j \ldots$ and so on). Without loss of generality we can assume that $\mathcal{D}_{\epsilon',j}$ does not depend on j and we shall use $\mathcal{D}_{\epsilon'}$ instead of $\mathcal{D}_{\epsilon',j}$. Let σ_j be a local parameter in $\mathcal{D}_{\epsilon,j}$ which gives the above-mentioned identification of $\mathcal{D}_{\epsilon,j}$ and $\mathcal{D}_{\epsilon}$ (that is, $\mathcal{D}_{\epsilon,j} = \{\sigma_j \in \mathbb{C}, |\sigma_j| < \epsilon\}$),

$\bar{\mathcal{D}}_j = \{\sigma_j \in \mathbb{C}, |\sigma_j| \le \epsilon_{2,j}\}$, $U = \Delta - \bigcup_{j=1}^{2} \bar{D}_j$, $W_{1U} = U \times \mathcal{D}_{\epsilon'}$,

$W_U = f^{-1}(U) \times \mathcal{D}_{\epsilon'}$, $F_U = (f|_{f^{-1}(U)}) \times (\text{identity}): W_U \longrightarrow W_{1U}$

and $h_U: W_U \longrightarrow \mathcal{D}_{\epsilon'}$, $p_U: W_{1U} \longrightarrow \mathcal{D}_{\epsilon'}$ are evident projections.

We have evident embeddings

$K_j \times \mathcal{D}_{\epsilon'} \longrightarrow U \times \mathcal{D}_{\epsilon'} (= W_{1U})$, $V'_{K_j} \times \mathcal{D}_{\epsilon'} \longrightarrow f^{-1}(U) \times \mathcal{D}_{\epsilon'} (= W_U)$

$(K_j = \{\sigma_j \in \mathbb{C} \mid \epsilon_{2,j} < |\sigma_j| < \epsilon_{1,j}, \epsilon_{1,j} < \epsilon, \epsilon_{2,j} > 0\})$ and

$h'_{K_j} = h_U|_{V'_{K_j} \times \mathcal{D}_{\epsilon'}}$, $p'_{K_j} = p_U|_{K_j \times \mathcal{D}_{\epsilon'}}$. Taking ϵ', $\epsilon_{1,j}$ smaller and $\epsilon_{2,j}$ greater we can suppose that i_{K_j} could be extended to an embedding $i_{\bar{K}_j}: \bar{K}_j \times \mathcal{D}_{\epsilon'} \longrightarrow \mathcal{D}_{\epsilon,j} \times \mathcal{D}_{\epsilon'}$

(where $\bar{K}_j = \{\sigma_j \in \mathbb{C} \,|\, \epsilon_2 \le |\sigma_j| \le \epsilon_1\}$ and that

$i_{\bar{K}_j}(\bar{K}_j \times \mathcal{D}_{\epsilon'}) \cap (0) \times \mathcal{D}_{\epsilon'} = \emptyset$. Now $(\mathcal{D}_{\epsilon,j} \times \mathcal{D}_{\epsilon'}) - i_{\bar{K}_j}(\bar{K}_j \times \mathcal{D}_{\epsilon'})$

has two connected components. Denote by U_j' one of them which does not contain $(0)\times \mathcal{D}_{\epsilon'}$. Let $W_{1,j}$ be the interior part of $(\mathcal{D}_{\epsilon,j}\times\mathcal{D}_{\epsilon'}) - U_j'$, $W^{(j)} = F_j'^{-1}(W_{1,j})$, $p_j = p_j'|_{W_{1,j}}$, $F_j = F_j'|_{W^{(j)}}: W^{(j)} \longrightarrow W_{1,j}$, $h_j = h_j'|_{W^{(j)}}$. We can consider i_{K_j} (corresp. $i_{V,j}$) as biholomorphic embeddings of $K_j\times\mathcal{D}_{\epsilon'}$ (corresp. $V_{K_j}'\times\mathcal{D}_{\epsilon'}$) into $W_{1,j}$ (corresp. $W^{(j)}$). Using the above-mentioned $K_j\times\mathcal{D}_{\epsilon'} \longrightarrow W_{1U}$ (corresp. $V_{K_j}'\times\mathcal{D}_{\epsilon'} \longrightarrow W_U$) we form new complex manifold W_1 (corresp. W) which is the union

$$W_{1U} \underset{\bigcup_{j=1}^{r} K_j\times D_{\epsilon'} \,=\, \bigcup_{j=1}^{r} i_{K_j}(K_j\times D_{\epsilon'})}{\smile} \left(\bigcup_{j=1}^{r} W_{1j}\right)$$

$$\text{(corresp. } W_U \underset{\bigcup_{j=1}^{r} V_{K_j}'\times D_{\epsilon'} \,=\, \bigcup_{j=1}^{r} i_{V,j}(V_{K_j}'\times D_{\epsilon'})}{\smile} \left(\bigcup_{j=1}^{r} W^{(j)}\right)\text{).}$$

Now we identify $W_{1U}, W_{11}, \cdots, W_{1r}$ (corresp. $W_U, W^{(1)}, \ldots, W^{(r)}$) with the corresponding open subsets of W_1 (corresp. W). We define $F: W \longrightarrow W_1$, $p: W_1 \longrightarrow \mathcal{D}_{\epsilon'}$, $h: W \longrightarrow \mathcal{D}_{\epsilon'}$ as follows:

$$F(x) = \begin{cases} F_U(x) & \text{if } x \in W_U \\ F_j(x) & \text{if } x \in W^{(j)} \end{cases},$$

$$p(y) = \begin{cases} p_U((y) & \text{if } y \in W_{1U} \\ p_j(y) & \text{if } y \in W_{1,j} \end{cases},$$

$$h(x) = \begin{matrix} h_U(x), \ x \in W_U, \\ h_j(x), \ x \in W^{(j)}. \end{matrix}$$

Using commutativity of (1) and (2) of Lemma 4 we can verify that F, p, and h are well defined holomorphic maps and that the diagram:

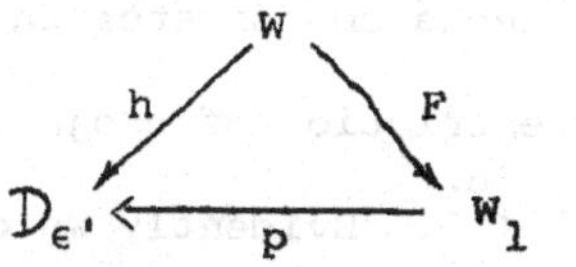

is commutative.

It is easy to see that this diagram satisfies all demands of Theorem 8. Q.E.D.

<u>Proof of Lemma 4</u>. Let $\mathcal{J}(\omega)$ be the modular function (the "absolute invariant"). Kodaira remarks in ([13], §7) that there exist $N_o > 0$ and a convergent power series $\mathfrak{P}(z)$ such that

$$\mathcal{J}(\omega) = e^{-2\pi i\omega} + \mathcal{P}(e^{2\pi i\omega}) \quad \text{for} \quad \text{Im}\,\omega > N_o.$$

Let $\epsilon > 0$ and $M_\epsilon = \{(\sigma,\tau) \in \mathbb{C}^2 \,|\, |\sigma| < \epsilon,\ |\tau| < \epsilon\}$. For ϵ sufficiently small we can define the following holomorphic function in M_ϵ

$$B(\sigma,\tau) = \sqrt{1 - \frac{\sigma^{n_1} - \tau^{n_1}}{1+(\sigma^{n_1}-\tau^{n_1})\mathcal{P}(\sigma^{n_1}-\tau^{n_1})}}, \quad (B(0,0)=1),$$

where the positive integer n_1 will be defined later.

Let X be a complex variety in $M_\epsilon \times \mathbb{C}P^2$ defined by the equation

$$z_o z_2^2 = 4z_1^3 - 3z_1 z_o^2 - B(\sigma,\tau)z_o^3$$

where $(z_o:z_1:z_2)$ are homogeneous coordinates in $\mathbb{C}P^2$, $\mathfrak{P}: X \longrightarrow M_\epsilon$ be defined by restriction of projection $M_\epsilon \times \mathbb{C}P^2 \longrightarrow M_\epsilon$. Let $z = \sigma^{n_1} - \tau^{n_1}$. Evidently we can write $B(\sigma,\tau) = B(z) = 1-zB_1(z)$, where $B_1(z)$ is a convergent power series and $B_1(0) \neq 0$. Let $\sigma_1 = \sigma\sqrt[n_1]{B_1(\sigma^{n_1}-\tau^{n_1})}$, $\tau_1 = \tau\sqrt[n_1]{B_1(\sigma^{n_1}-\tau^{n_1})}$ where we fix a branch of $\sqrt[n_1]{B_1(z)}$, and

$M_{1\tilde{\epsilon}} = \{(\sigma_1,\tau_1) \in \mathbb{C}^2 \,|\, |\sigma_1| < \tilde{\epsilon},\ |\tau_1| < \tilde{\epsilon}\}$, $\tilde{\epsilon}$ is a sufficiently small positive number.

Taking ϵ smaller we can write the equation of X as follows:

$$z_o z_2^2 = 4z_1^3 - 3z_1 z_o^2 + (-1 + \sigma_1^{n_1} - \tau_1^{n_1})z_o^3 .$$

For any $t \in M_\epsilon$ $\mathfrak{F}^{-1}(t)$ is a cubic curve in $\mathbb{C}P^2$ which has no singularities on the line $z_o = 0$. $\mathfrak{F}^{-1}(t)$ is singular if and only if $(\sigma_1(t))^{n_1} - (\tau_1(t))^{n_1} = 0$ (or equivalently $(\sigma(t))^{n_1} - (\tau(t))^{n_1} = 0$).

Let $M'_\epsilon = \{(\sigma,\tau) \in M_\epsilon \,|\, \sigma^{n_1} - \tau^{n_1} \neq 0\}$, $x = \frac{z_1}{z_o}$, $y = \frac{z_2}{z_o}$.

Taking ϵ smaller we can assume that there exists $r > 0$ such that the equation (on x) $4x^3 - 3x - 1 + \sigma_1^{n_1} - \tau_1^{n_1} = 0$ has exactly two roots $x_1(\sigma,\tau), x_2(\sigma,\tau)$ with $|x_i(\sigma,\tau) + \frac{1}{2}| < r$ when $(\sigma,\tau) \in M'_\epsilon$. Let $D(-\frac{1}{2}, r) = \{x \in \mathbb{C} \,|\, |x + \frac{1}{2}| < r\}$ and $\gamma(\sigma,\tau)$ be a segment of the straight (real) line in $D(-\frac{1}{2}, r)$ connecting $x_1(\sigma,\tau)$ and $x_2(\sigma,\tau)$,

$$\delta_1(\sigma,\tau) = \left\{(x,y) \in \mathbb{C}^2 \,\middle|\, x \in \gamma(\sigma,\tau),\ y^2 = 4x^3 - 3x - 1 + \sigma_1^{n_1} - \tau_1^{n_1}\right\},$$

$(\sigma,\tau) \in M'_\epsilon$. $\delta_1(\sigma,\tau)$ is a homologically non-trivial 1-dimensional cycle in $\mathfrak{F}^{-1}(\sigma,\tau)$. Fixing an orientation of $\delta_1(\sigma_o,\tau_o)$ for some $(\sigma_o,\tau_o) \in M'_\epsilon$ we can check that for any close path γ in M'_ϵ with origin (σ_o,τ_o), $\delta_1(\sigma,\tau)$ has continuous changing along γ and returns to $\delta_1(\sigma_o,\tau_o)$ with the same orientation. Now we can speak about a continuous family of oriented 1-dimensional cycles $\{\delta_1(\sigma,\tau) \in \mathfrak{F}^{-1}(\sigma,\tau),\ (\sigma,\tau) \in M'_\epsilon\}$.

Let $\alpha(\sigma,\tau)$ be a holomorphic differential on $\mathfrak{F}^{-1}(\sigma,\tau)$, $(\sigma,\tau) \in M'_\epsilon$, which in the part of $\mathbb{C}P^2$ with $z_o \neq 0$ is given by

$\alpha(\sigma,\tau) = \frac{dx}{Y}$, $\omega_1(\sigma,\tau) = \int_{\delta_1(\sigma,\tau)} \alpha(\sigma,\tau)$. $\omega_1(\sigma,\tau)$ is a single-valued holomorphic function in M'_ϵ without zeroes. (If $\omega_1(\sigma',\tau') = 0$ for some $(\sigma',\tau') \in M'_\epsilon$ we would have $\int_{\delta_1(\sigma',\tau')} \alpha(\sigma',\tau') = 0$, $\int_{\delta_1(\sigma',\tau')} \bar{\alpha}(\sigma',\tau') = 0$ and $\delta_1(\sigma',\tau')$ is homologically zero).

Let $\mathcal{V}$ be a small open neighborhood of (σ_o,τ_o) in M'_ϵ, $\{\delta_2(\sigma,\tau) \in \mathfrak{F}^{-1}(\sigma,\tau), (\sigma,\tau)\in \mathcal{V}\}$ be a continuous family of oriented 1-cycles such that for any $(\sigma,\tau) \in \mathcal{V}$, $\delta_1(\sigma,\tau),\delta_2(\sigma,\tau)$ generate $H_1(\mathfrak{F}^{-1}(\sigma,\tau),\mathbb{Z})$ and $\operatorname{Im}\left[\int_{\delta_2(\sigma,\tau)} \alpha(\sigma,\tau)/\omega_1(\sigma,\tau\right] > 0$. We denote $\omega(\sigma,\tau) = \int_{\delta_2(\sigma,\tau)} \alpha/\omega_1(\sigma,\tau)$ and get by standard arguments that it is a multi-valued holomorphic function in M'_ϵ such that for any close path γ, in M'_ϵ, $\gamma \ni (\sigma_o,\tau_o)$ analytic continuation of $\omega(\sigma,\tau)$ along γ gives $\omega(\sigma_o,\tau_o)+N$ from $\omega(\sigma_o,\tau_o)$, where $N \in \mathbb{Z}$.

If $t = (\sigma,\tau) \in M'_\epsilon$ then $\mathfrak{F}^{-1}(t)$ is given in $\mathbb{C}P^2$ by the following equation: $z_o z_2^2 = 4z_1^3-3z_1z_o^2-B(\sigma,\tau)z_o^3$ or in non-homogeneous coordinates,

$$x = \frac{z_1}{z_o}, \quad y = \frac{z_2}{z_o}: \quad y^2 = 4x^3 - 3x - B(\sigma,\tau).$$

We get that absolute invariant of $\mathfrak{F}^{-1}(t)$

$$\mathcal{J}(t)\left(=\frac{g_2^3}{g_2^3-27g_3^2}\right)=\frac{27}{27-27B^2(\sigma,\tau)}=\frac{1}{1-B^2(\sigma,\tau)}=\frac{1+(\sigma^{n_1}-\tau^{n_1})\mathfrak{P}(\sigma^{n_1}-\tau^{n_1})}{\sigma^{n_1}-\tau^{n_1}}$$

$$=\frac{1}{\sigma^{n_1}-\tau^{n_1}}+\mathfrak{P}(\sigma^{n_1}-\tau^{n_1}).$$

Fix $\alpha \in \mathbb{C}^*$, $\alpha^{n_1} \neq 1$, and denote

$$\ell_\alpha = \{(\sigma,\tau) \in M_\epsilon \mid \sigma = \alpha\tau\}, \quad Z_\alpha = \mathfrak{F}^{-1}(\ell_\alpha).$$

We see that $\mathcal{J}(t)|_{\ell_\alpha}$ has in $(0,0)$ a pole of order n_1. From results of Kodaira it then follows that if $\tilde{Z}_\alpha$ is the minimal desingularization of Z_α then $\tilde{Z}_\alpha$ has over $(0,0)$ a singular fiber of type I_{n_1} and

$$\omega(\alpha\tau,\tau) = \frac{1}{2\pi i}\, n_1 \log \tau + f(\tau), \quad 0 < |\tau| < \tilde{\epsilon},$$

where $\tilde{\epsilon}$ is sufficiently small and $f(\tau)$ is a holomorphic function in $0 \leq |\tau| < \tilde{\epsilon}$. Taking $\tilde{\epsilon}$ smaller we can assume that $\operatorname{Im} \omega(\alpha\tau,\tau) > N_0$ (for $0 < |\tau| < \tilde{\epsilon}$) and thus

$$\mathcal{J}(\alpha\tau,\tau) = e^{-2\pi i\omega(\alpha\tau,\tau)} + \mathfrak{P}(e^{2\pi i\omega(\alpha\tau,\tau)}) \quad (\text{for } 0 < |\tau| < \tilde{\epsilon}).$$

Denote $q(\sigma,\tau) = e^{2\pi i\omega(\sigma,\tau)}$, $p(\sigma,\tau) = \sigma^{n_1} - \tau^{n_1}$. We have $q(\alpha\tau,\tau) = \tau^{n_1}e^{2\pi i f(\tau)}$ $(0 < |\tau| < \tilde{\epsilon})$, that is, $q(\alpha\tau,\tau)$ is a holomorphic function in $0 \leq |\tau| < \tilde{\epsilon}$. Let $\dot{D}_{\tilde{\epsilon}} = \{\tau \in \mathbb{C} \mid 0 < |\tau| < \tilde{\epsilon}\}$ and $\varphi_\alpha : \dot{D}_{\tilde{\epsilon}} \longrightarrow \mathbb{C}^2$ be a holomorphic map defined by

$\tau \longrightarrow (q(\alpha\tau,\tau),p(\alpha\tau,\tau))$. Since

$$\frac{1}{q(\alpha\tau,\tau)} + \mathfrak{P}(q(\alpha\tau,\tau)) = \mathcal{J}(\alpha\tau,\tau) = \frac{1}{p(\alpha\tau,\tau)} + \mathfrak{P}(p(\alpha\tau,\tau))$$

we have that $\Psi_\alpha(\dot{D}_{\tilde{\epsilon}})$ is contained in the subset C_α of $\mathbb{C}^2$ given by $C_\alpha = \{(p,q) \in \mathbb{C}^2 \mid p + pq\mathfrak{P}(q) = q+pq\mathfrak{P}(p),\ |p| < \tilde{\epsilon}_1,\ |q| < \tilde{\epsilon}_1\}$ where we suppose that $\tilde{\epsilon}$ and $\tilde{\epsilon}_1$ are sufficiently small. But for $\tilde{\epsilon}_1$ sufficiently small C_α is a non-singular analytic curve and because it evidently contains the set

$$\Delta = \{(p,q) \in \mathbb{C}^2 \mid p = q,\ |p| < \tilde{\epsilon}_1,\ |q| < \tilde{\epsilon}_1\}$$

we have $C_\alpha = \Delta$. We see that $p(\alpha\tau,\tau) = q(\alpha\tau,\tau)$ for $0 < |\tau| < \tilde{\epsilon}$. Hence

$$q(\sigma,\tau)\big|_{\ell_\alpha} = (\sigma^{n_1}-\tau^{n_1})\big|_{\ell_\alpha} \quad \text{and} \quad q(\sigma,\tau) = \sigma^{n_1}-\tau^{n_1} \quad \text{in } D_\epsilon.$$

Let

$$\Sigma_\epsilon = \{(\sigma,\tau) \in D_\epsilon \mid \sigma^{n_1}-\tau^{n_1} = 0\},\ \dot{\Sigma}_\epsilon = \Sigma_\epsilon-(0,0),$$
$$\dot{M}_\epsilon = M_\epsilon-(0,0),\ M = \dot{M}_\epsilon \times \mathbb{C}^*,\ S = \dot{\Sigma}_\epsilon \times \mathbb{C}^*, \qquad \mathcal{D}$$

be a cyclic group of analytic automorphisms of $M' = M-S$ generated by the transformation $(\sigma,\tau,w) \longrightarrow (\sigma,\tau,(\sigma^{n_1}-\tau^{n_1})w)$. Using Kodaira's remark in ([13], p. 597) we construct from M and $\mathcal{D}$ a new complex manifold which we call Kodaira factor-space and denote by $M/\mathcal{D}$ (holomorphic map Ψ: $M \longrightarrow M/\mathcal{D}$ is given by: $\Psi(z') = \Psi(z'')$, $z',z'' \in M$, if and only if either $z',z'' \in M'$ and

$z' = g'(z'')$ for some $g' \in \mathcal{D}$ or $z' = z''$). (It is easy to verify that the condition for existence of $M/\mathcal{D}$ indicated by Kodaira in [13] is satisfied in our case).

Let $P_0(\sigma,\tau) \in \mathcal{F}^{-1}((\sigma,\tau))$ be given by

$$z_0(P_0(\sigma,\tau)) = 0, \quad z_1(P_0(\sigma,\tau)) = 0, \quad z_2(P_0(\sigma,\tau)) = 1,$$

$X_{(0)} = \{(\sigma,\tau,z_0,z_1,z_2) \in X \,|\, (\sigma,\tau) \in M_\epsilon^{\cdot}$ and (z_0,z_1,z_2) is a non-singular point of $\mathcal{F}^{-1}(\sigma,\tau)\}$, $\tilde{C}_t = \mathcal{F}^{-1}(t) \cap X_{(0)}$, $t = (\sigma,\tau) \in M_\epsilon^{\cdot}$. Evidently we can consider $\alpha(\sigma,\tau)$ as a holomorphic differential on $\tilde{C}_t$ for any $t \in M_\epsilon^{\cdot}$. Take $(\sigma',\tau') \in \Sigma_\epsilon^{\cdot}$. We can define a 1-cycle $\delta_1 \subset (\sigma',\tau')$ in $\tilde{C}(\sigma',\tau')$ such that $\delta_1(\sigma',\tau')$ could be connected with the previously defined $\delta_1(\sigma_0,\tau_0)$ by continuous family of 1-cycles in $\mathcal{F}^{-1}(t)$, $t \in \gamma$, where γ is a path in $M_\epsilon^{\cdot}$ connecting (σ_0,τ_0) and (σ',τ') and $\mathrm{Int}(\gamma) \subset M_\epsilon^{\cdot} - \Sigma_\epsilon^{\cdot}$. A direct verification then shows that

$$\int_{\delta_1(\sigma',\tau')} \alpha(\sigma',\tau') \neq 0.$$

We see that $\omega_1(\sigma,\tau)$ is actually defined in $M_\epsilon^{\cdot}$ and for each $(\sigma,\tau) \in M_\epsilon^{\cdot}$, $\omega_1(\sigma,\tau) \neq 0$. (This proves also that $\omega_1(\sigma,\tau)$ is a (single-valued) holomorphic function in M_ϵ and $\omega_1(\sigma,\tau) \neq 0$, $(\sigma,\tau) \in M_\epsilon$).

Let $\alpha_1(\sigma,\tau) = \frac{1}{\omega_1(\sigma,\tau)}\alpha(\sigma,\tau), ((\sigma,\tau) \in M_\epsilon^{\cdot})$. Using $e^{2\pi i\omega(\sigma,\tau)} = \sigma^{n_1} - \tau^{n_1}$ we define a holomorphic map

$$\Psi_0 : X_{(0)} \longrightarrow M/\mathcal{D} \text{ by } \Psi_0(a) = [\sigma(a),\tau(a);\, \exp(2\pi i \int_{P_0(\sigma(a),\tau(a))} \alpha_1(\sigma(a),\tau(a)),$$

where $\int_{P_o(\sigma(a),\tau(a))}^{a}$ is taken along a path in $\tilde{C}_{(\sigma(a),\tau(a))}$

$(\sigma(a) = \sigma(\mathfrak{F}'(a)), \tau(a) = \tau(\mathfrak{F}'(a)))$. It is clear that Ψ_o is onto and a 1-1 map. Hence Ψ_o is an isomorphism of complex manifolds. We shall identify $X_{(o)}$ and $M/\mathcal{D}$. Kodaira's construction of logarithmic transform (see [14], p. 770) shows that a certain residue class $k \bmod m$ is defined for a multiple fiber of type ${}_mI_b$, $b \geq 1$, k is relatively prime to m. (Actually a topological definition for such a k could be given. For us it will be important only that Kodaira's construction shows the analytical uniqueness of a neighborhood of a multiple fiber ${}_mI_b$, $b \geq 1$, when k is fixed).

Take the corresponding $k = k(f')$ for the family $f'\colon V' \longrightarrow D_\epsilon$ given in our Lemma and let $k' \in \mathbb{Z}$ be such that $kk' \equiv 1 \pmod m$. Let $d = \text{g.c.d.}\ (k',b)$, $k' = k_1'd$, $b = b_1d$, $n = mb$, $n_1 = mb_1$. (That is the definition of n_1).

Denote $\rho_1 = \exp(2\pi i \frac{k_1'}{n_1})$. For $j = 1,\cdots,n_1-1$, let $x_j(\sigma,\tau)$, $y_j(\sigma,\tau)$ be two holomorphic functions in M_ϵ^{*} defined as follows:

$$x_j(\sigma,\tau) = \frac{z_1([\sigma,\tau;w_j(\sigma,\tau)])}{z_o([\sigma,\tau;w_j(\sigma,\tau)])}, \quad y_j(\sigma,\tau) = \frac{z_2([\sigma,\tau;w_j(\sigma,\tau)])}{z_o([\sigma,\tau;w_j(\sigma,\tau)])}$$

where

$$w_j(\sigma,\tau) = \begin{cases} \prod_{\ell=1}^{j} (\sigma-\rho_1^{\ell}\tau)^{-1} & \text{if } \sigma-\rho_1^{\ell}\tau \neq 0, \quad \ell = 1,\cdots,j, \\ \prod_{\ell=j+1}^{n_1} (\sigma-\rho_1^{\ell}\tau) & \text{if for some } \ell' \in (1,2,\cdots,j), \\ & \sigma-\rho_1^{\ell'}\tau = 0 . \end{cases}$$

We use the following remarks:

$$X_{(o)} \cap (z_o=0) = \bigcup_{(\sigma,\tau)\in M_\epsilon^{\cdot}} P_o(\sigma,\tau), \quad \Psi_o(P_o(\sigma,\tau)) = [\sigma,\tau;1]$$

and for $\epsilon < \frac{1}{3}$

$$\left|\prod_{\ell=1}^{j}(\sigma-\rho_1^{\ell}\tau)\right| < 1, \qquad \left|\prod_{\ell=j+1}^{n_1}(\sigma-\rho_1^{\ell}\tau)\right| < 1$$

(that is, for $(\sigma,\tau) \in \Sigma^{\cdot}$, $w_j(\sigma,\tau) \neq 1$, $z_o([\sigma,\tau;w_j(\sigma,\tau)]) \neq 0$, and for $(\sigma,\tau) \in M_\epsilon^{\cdot}-\Sigma^{\cdot}$ an equality

$$\prod_{\ell=j+1}^{n_1}(\sigma-\rho_1^{\ell}\tau) = (\sigma^{n_1}-\tau^{n_1})^N, \quad N \in \mathbb{Z} ,$$

gives first $N > 0$ and then

$$1 = \left|\frac{(\sigma^{n_1}-\tau^{n_1})^N}{\prod_{\ell=j+1}^{n_1}(\sigma-\rho_1^{\ell}\tau)}\right| = \left|\sigma^{n_1}-\tau^{n_1}\right|^{N-1}\left|\prod_{\ell=1}^{j}(\sigma-\rho_1^{\ell}\tau)\right| < 1.$$

Contradiction).

Since $M_\epsilon - \dot{M}_\epsilon = (0,0)$ we see that $x_j(\sigma,\tau), y_j(\sigma,\tau)$ are holomorphic functions defined in M_ϵ.

For $(\sigma',\tau') \in \Sigma_\epsilon$, $\mathfrak{F}^{-1}(\sigma',\tau')$ is the same rational plane curve, say C, given in $\mathbb{C}P^2$ by the equation

$$z_o z_2^2 = 4z_1^3 - 3z_1 z_o^2 - z_o^3.$$

Hence we can assume that $\delta_1(\sigma',\tau')$ chosen above for $(\sigma',\tau') \in \dot{\Sigma}_\epsilon$ is the same 1-cycle on all $\mathfrak{F}^{-1}(\sigma',\tau')$, $(\sigma',\tau') \in \dot{\Sigma}_\epsilon$ (that is, images of $\delta_1(\sigma',\tau')$ by $M_\epsilon \times \mathbb{C}P^2 \longrightarrow \mathbb{C}P^2$ coincide with some 1-cycle δ_{1C} on the non-singular part $\tilde{C}$ of C). Now define by

$$x_C = \left.\frac{z_1}{z_o}\right|_C, \quad y_C = \left.\frac{z_2}{z_o}\right|_C, \quad \alpha_C = \frac{dx_C}{y_C}, \quad \omega_{1C} = \int_{\delta_{1C}} \alpha_C,$$

$$\alpha_{1C} = \frac{1}{\omega_{1C}}\alpha_C, \qquad p_o = (z_o=0, z_1=0, z_2=1) \in C,$$

$$w_C(p) = e^{2\pi i \int_{p_o}^{p} \alpha_{1C}}$$

where $p \in \tilde{C}$ and $\int_{p_o}^{p}$ is taken along some path in $\tilde{C}$,

$T = \dfrac{y_C}{x_C + \frac{1}{2}}$ (that is, $x_C = \frac{T^2}{4}+1$, $y_C = \frac{T(T^2+6)}{4}$). A direct calculation shows that (changing if necessary the orientation of δ_{1C}) we have

(1) $\tau = i\sqrt{6}\,\frac{w_c+1}{w_c-1}$, $x_c = \frac{-3w_c}{4(w_c-1)^2} - \frac{1}{2}$, $y_c = \frac{-3i\sqrt{6}\,w_c(w_c+1)}{4(w_c-1)^3}$.

Take $\ell' \in (1,2,\cdots,n_1)$ and let

$$\Lambda_{\ell'} = \{(\sigma,\tau) \in M_\epsilon \,|\, \sigma - \rho_1^{\ell'}\tau = 0\}, \quad \dot{\Lambda}_{\ell'} = \Lambda_{\ell'} - (0,0),$$

$$x_{j,\ell'} = x_j(\sigma,\tau)\big|_{\Lambda_{\ell'}}, \quad y_{j,\ell'} = y_j(\sigma,\tau)\big|_{\Lambda_{\ell'}}, \quad w_{j\ell'} = w_j(\sigma,\tau)\big|_{\Lambda_{\ell'}}.$$

Since $w_{j\ell'} \longrightarrow 0$ or ∞ when $(\sigma,\tau) \longrightarrow (0,0)((\sigma,\tau) \in \Lambda_{\ell'})$ we see from (1) that

$$\lim_{\substack{(\sigma,\tau)\to(0,0)\\(\sigma,\tau)\in\Lambda_{\ell'}}} x_{j,\ell'} = -\frac{1}{2}, \qquad \lim_{\substack{(\sigma,\tau)\to(0,0)\\(\sigma,\tau)\in\Lambda_{\ell'}}} y_{j,\ell'} = 0 .$$

This shows that $x_j(0,0) = -\frac{1}{2}$, $y_j(0,0) = 0$.

Now let $\varphi(x)$ be a (single-valued) holomorphic function defined in some neighborhood of $x = -\frac{1}{2}$ by $\varphi(x) = 2\sqrt{x-1}$ and $\varphi(-\frac{1}{2}) = i\sqrt{6}$. Let $y' = y + \varphi(x)(x+\frac{1}{2})$, $x' = y-\varphi(x)(x+\frac{1}{2})$ $(y,x'$ are some local parameters at the point $(x = -\frac{1}{2}, y = 0)$ of $\mathbb{C}^2)$. From $x_j(0,0) = -\frac{1}{2}$, $y_j(0,0) = 0$ we see that (taking ϵ smaller if necessary) the following holomorphic functions in M_ϵ are well defined:

$$y_j'(\sigma,\tau) = y_j(\sigma,\tau) + \varphi(x_j(\sigma,\tau))(x_j(\sigma,\tau) + \tfrac{1}{2}),$$

$$x_j'(\sigma,\tau) = y_j(\sigma,\tau) - \varphi(x_j(\sigma,\tau))(x_j(\sigma,\tau) + \tfrac{1}{2}).$$

Because

$$y_j^2(\sigma,\tau) = 4x_j^3(\sigma,\tau) - 3x_j(\sigma,\tau) - B(\sigma,\tau)$$

we have

$$y_j'(\sigma,\tau)x_j'(\sigma,\tau) = (\sigma^{n_1}-\tau^{n_1})e(\sigma,\tau)$$

where

$$e(\sigma,\tau) = \frac{B(\sigma,\tau)-1}{\sigma^{n_1}-\tau^{n_1}} ,$$

that is, $e(\sigma,\tau)$ is holomorphic in $\mathcal{M}_\epsilon$ and $e(0,0) \neq 0$.

Define on $\Lambda_{\ell'}$, $T_{j,\ell'}(\sigma,\tau) = \dfrac{y_{j,\ell'}(\sigma,\tau)}{x_{j,\ell'}(\sigma,\tau)+\frac{1}{2}}$. From (1) we see that on $\Lambda_{\ell'}$

$$T_{j,\ell'}(\sigma,\tau) = i\sqrt{6}\,\frac{w_{j,\ell'}(\sigma,\tau)+1}{w_{j,\ell'}(\sigma,\tau)-1} .$$

For $\ell' \in (1,2,\cdots,j)$, we have

$$w_{j,\ell'}(\sigma,\tau) = \prod_{\ell=j+1}^{n_1} \sigma-\rho_1^{\ell}\tau).$$

Thus

$$\lim_{\Lambda_{\ell'}\ni(\sigma,\tau)\to(0,0)} w_{j,\ell'}(\sigma,\tau) = 0 \quad\text{and}\quad \lim_{\Lambda_{\ell'}\ni(\sigma,\tau)\to(0,0)} T_{j,\ell'}(\sigma,\tau) = -i$$

Since $T^2_{j,\ell'}(\sigma,\tau) = [\varphi(x_{j,\ell'}(\sigma,\tau)]^2$ (we use $T^2 = \dfrac{y_C^2}{(x_C+\frac{1}{2})^2} = 4(x_C-1)$)

and $\varphi(-\frac{1}{2}) = i\sqrt{6}$ we have $\varphi(x_{j,\ell'}(\sigma,\tau)) = -T_{j,\ell'}(\sigma,\tau)$ and

$y_j'(\sigma,\tau)\big|_{\Lambda_{\ell'}} = y_{j,\ell'} - T_{j,\ell'}(x_{j,\ell'} + \frac{1}{2}) \equiv 0.$ We obtain

$y_j'(\sigma,\tau) = \prod_{\ell=1}^{j}(\sigma-\rho_1^{\ell}\tau))e_j'(\sigma,\tau)$ where $e_j'(\sigma,\tau)$ is holomorphic in M_ϵ.

Now let $\ell' \in (j+1,\cdots,n_1)$. We have

$$w_{j,\ell'}(\sigma,\tau) = \prod_{\ell=1}^{j}(\sigma-\rho_1^{\ell}\tau)^{-1}. \text{ Thus}$$

$$\lim_{\Lambda_{\ell'}\ni(\sigma,\tau)\to(0,0)} w_{j,\ell'}(\sigma,\tau) = \infty \quad \text{and} \quad \lim_{\Lambda_{\ell'}\ni(\sigma,\tau)\to(0,0)} T_{j,\ell'}(\sigma,\tau) = i\sqrt{6}.$$

We have $\varphi_{x_j',\ell'}(\sigma,\tau) = T_{j,\ell'}(\sigma,\tau)$ and

$$x_j'(\sigma,\tau)\Big|_{\Lambda_{\ell'}} = y_{j,\ell'} - T_{j,\ell'}(x_{j,\ell'} + \tfrac{1}{2}) = 0.$$

We obtain

$$x_j'(\sigma,\tau) = \prod_{\ell=j+1}^{n_1}(\sigma-\rho_1^{\ell}\tau))e_j''(\sigma,\tau)$$

where $e_j''(\sigma,\tau)$ is holomorphic in M_ϵ. Now

$$(\sigma^{n_1}-\tau^{n_1})e(\sigma,\tau) = y_j'(\sigma,\tau)x_j'(\sigma,\tau) = \prod_{\ell=1}^{n_1}(\sigma-\rho_1^{\ell}\tau))e_j'(\sigma,\tau)e_j''(\sigma,\tau).$$

Hence $e(\sigma,\tau) = e_j'(\sigma,\tau)e_j''(\sigma,\tau)$ and $e_j'(0,0) \neq 0$, $e_j''(0,0) \neq 0$.

X has only one singular point

$$\mathfrak{o} = (z_0(\mathfrak{o}) = 1,\ z_1(\mathfrak{o}) = -\tfrac{1}{2};\ z_2(\mathfrak{o}) = 0;\ \sigma(\mathfrak{o}) = 0,\ \tau(\mathfrak{o}) = 0).$$

We can find an open neighborhood U of $\bullet$ in $M_\epsilon \times \mathbb{C}P^2$ such that (σ,τ,x',y') are local coordinates in U $\left(x' = y-\varphi(x)(x+\frac{1}{2}),\ y' = y+\varphi(x)(x+\frac{1}{2}),\ x = \frac{z_1}{z_0},\ y = \frac{z_2}{z_0}\right)$. $X_U = X \cap U$ is given in U by the equation $x'y' = (\sigma^{n_1}-\tau^{n_1})e(\sigma,\tau)$.

Now we take n_1-1 copies of $\mathbb{C}P^1$, say $\mathbb{C}P^1_i$, $i = 1,2,\cdots,n_1-1$, with homogeneous coordinates $(\xi_{0i} : \xi_{1i})$, $i = 1,\cdots,n_1-1$, and consider in $U \times \prod_{i=1}^{n_1-1} \mathbb{C}P^1_i$ a subvariety Y_U defined by the following system of equations:

$$y'\xi_{01} = \xi_{11}(\sigma-\rho_1\tau)e(\sigma,\tau)$$

$$\xi_{1i}\xi_{0i+1} = \xi_{0i}\xi_{1i+1}(\sigma-\rho_1^{i+1}\tau),\quad i = 1,\cdots,n_1-2,$$

$$x'\xi_{1n_1-1} = \xi_{0n_1-1}(\sigma-\tau).$$

It is easy to see that $pr_U(Y_U) = X_U$. Let $s\colon Y_U \longrightarrow X_U$ be the map induced by projection, $X_{U,\tau} = \{a \in X_U,\ \tau(\mathcal{F}(a)) = \tau\}$, $Y_{U,\tau} = s^{-1}(X_{U,\tau})$. Standard verification shows that Y_U is non-singular, $s|_{s^{-1}(X_U-\bullet)}\colon s^{-1}(X_U-\bullet) \longrightarrow X_U-\bullet$ is an isomorphism, $\tau\circ s$ is a holomorphic function on Y_U without critical values, and $s|_{Y_{U,0}}\colon Y_{U,0} \longrightarrow X_{U,0}$ is a minimal resolution of singularities of $X_{U,0}$ (that is, there are no exceptional curves of first kind of $Y_{U,0}$ in $s^{-1}(\bullet)$).

Now identifying $s_U^{-1}(X - \Theta)$ with $X_U - \alpha$ we get from $X-\Theta$ and Y_U a new complex manifold Y with a holomorphic map $\tilde{s}: Y \longrightarrow X$ such that (i) $\tilde{s}|_{\tilde{s}^{-1}(X-\Theta)}: \tilde{s}^{-1}(X-\Theta) \longrightarrow X-\Theta$ is an isomorphism, (ii) $\tau \cdot \tilde{s}$ is a holomorphic function on Y without critical values and (iii) if $X_\tau = \{a \in X, \tau(\tilde{\mathcal{P}}(a)) = \tau\}$, $Y_\tau = \tilde{s}^{-1}(X_\tau)$, then $\tilde{s}|_{Y_0}: Y_0 \longrightarrow X_0$ is a minimal resolution of singularities of X_0.

Let $\tilde{\mathcal{P}}' = \tilde{\mathcal{P}} \cdot \tilde{s}$, $\tilde{\mathcal{P}}'_{\tau'} = \tilde{\mathcal{P}}'|_{Y_{\tau'}}: Y_{\tau'} \longrightarrow D_{\epsilon,\tau'}$ where $D_{\epsilon,\tau'} = \{(\sigma,\tau) \in M_\epsilon | \tau = \tau'\}$. For any $a \in D_{\epsilon,\tau'} - D_{\epsilon,\tau'} \cap \Sigma_\epsilon$, $\tilde{\mathcal{P}}'^{-1}(a)$ is a non-singular elliptic curve (and a regular fiber of $\tilde{\mathcal{P}}'_{\tau'}$), $\tilde{\mathcal{P}}'^{-1}_{\tau'}((\rho_1^\ell \tau', \tau'))$, $\tau' \neq 0$, $\ell = 0,1,\cdots,n_1-1$, are singular fibers of $\tilde{\mathcal{P}}'_{\tau'}$ which have type I_1 and $\tilde{\mathcal{P}}'^{-1}_0((0,0))$ is a singular fiber of $\tilde{\mathcal{P}}'_0: Y_0 \longrightarrow D_{\epsilon,0}$ which is of type I_{n_1}.

Denote by $c_0^{(0)}$ the closure of $\tilde{s}^{-1}(\tilde{\mathcal{P}}^{-1}(0,0)-\Theta)$ in Y. For $j = 1,\cdots,m-1$, let $c_0^{(j)}$ be an algebraic curve in Y_U defined by the following system of equations:

$$\xi_{1i} = 0, \quad i = 1,\cdots,j-1, \quad \xi_{0i'} = 0, \quad i' = j+1,\cdots,n_1-1,$$

$$x' = y' = \sigma = \tau = 0.$$

It is clear that $c_0^{(0)}, c_0^{(1)}, \cdots, c_0^{(n_1-1)}$ are all the irreducible components of $\tilde{\mathcal{P}}'^{-1}(0,0)$. We assume $c_0^{(n_1)} = c_0^{(0)}$ and let $q_0^{(i)} = c_0^{(i)} \cap c_0^{(i+1)}$, $\tilde{c}^{(i)} = c_0^{(i)} - q_0^{(i)} - q_0^{(i-1)}$, $i = 0,1,\cdots,n_1-1$,

$$\tilde{C}_o = \bigcup_{i=0}^{n_1-1} \tilde{C}_o^{(i)}.$$

Take a point $a \in \tilde{C}_o^{(j)}$, $j \in (1,2,\cdots,n_1-1)$. We have $\xi_{1j}(a) \neq 0$, $\xi_{0j}(a) \neq 0$. Let U_a be an open neighborhood of a in Y_U ($\subset Y$) such that $\forall a' \in U_a$, $\xi_{1j}(a') \neq 0$, $\xi_{0j}(a') \neq 0$. We have a holomorphic function in U_a, $u_j = (\frac{\xi_{1j}}{\xi_{0j}})\Big|_{Y_U}$. Our differential $\alpha(\sigma,\tau) = \frac{dx}{y}$ which is defined on each $\tilde{C}_t$, $t \neq (0,0)$ can be rewritten in U_a by the following way:

First we have $y = \frac{y'+x'}{2}$, $y'-x' = 2\varphi(x)(x+\frac{1}{2})$. Since $\varphi(-\frac{1}{2}) \neq 0$ there exists a holomorphic function $\psi(z)$ defined in some neighborhood of $z = 0$ such that $\psi(0) \neq 0$ and $x + \frac{1}{2} = \frac{1}{2}(y'-x')\psi(y'-x')$. Thus we obtain

$$\alpha = \frac{dx}{y} = \frac{[\psi(y'-x')+(y'-x')\psi'(y'-x')]}{y'+x'}(dy'-dx').$$

Because differentiation here is along the fibers of $\mathfrak{F}'$, that is, when σ and τ are fixed, and on Y_U

$$y'x' = (\sigma^{n_1}-\tau^{n_1})e(\sigma,\tau)$$

we have $x'dy' + y'dx' = 0$. Hence

$$\alpha = [\psi(y'-x')+(y'-x')\psi'(y'-x')]\frac{dy'}{y'}.$$

But in U_a, $y' = u_j \prod_{\ell=1}^{j}(\sigma-\rho_1^{\ell}\tau)e(\sigma,\tau)$ and we get $\frac{dy'}{y'} = \frac{du_j}{u_j}$ and $\alpha = [\psi(y'-x')+(y'-x')\psi'(y'-x')]\frac{du_j}{u_j}.$

This formula shows that we can extend $\alpha(\sigma,\tau)$ also to $\tilde{C}_j$ where it is equal to $\psi(0)\frac{du_j}{u_j}$.

Let $\delta_{10}^{(j)}$ be a 1-cycle on $\tilde{C}_0^{(j)}$ defined by $|u_j| = 1$. A direct verification then shows that we can include $\delta_{10}^{(j)}$ in a continuous family of 1-cycles $\{\delta_1^{(j)}(\sigma,\tau), (\sigma,\tau) \in M_\epsilon\}$, $\delta_1^{(j)}(\sigma,\tau) \in \tilde{C}(\sigma,\tau)$, such that for $(\sigma,\tau) \neq 0$, $\delta_1^{(j)}(\sigma,\tau)$ is homologically equivalent in $\tilde{C}_{(\sigma,\tau)}$ to the above constructed $\delta_1(\sigma,\tau)$. Because

$$\omega_1(0,0) = \int_{\delta_{10}^{(j)}} \alpha(0,0) = \int_{|u_j|=1} \psi(0)\frac{du_j}{u_j} \neq 0$$

we see that we can extend $\alpha_1((\sigma,\tau)(\sigma,\tau) \neq 0)$ constructed above also to the non-singular part of $\mathfrak{F}'^{-1}(0,0)$ such that we get a holomorphic family $\{\alpha_1(\sigma,\tau), (\sigma,\tau) \in M_\epsilon\}$, where $\alpha_1(\sigma,\tau)$ is a holomorphic differential on $\tilde{C}_{(\sigma,\tau)}$ $(\tilde{C}_{(0,0)} = \tilde{C}_0)$.

Let $\tilde{P}^{(j)}$, $j = 1,\cdots,n_1-1$, be an analytic surface in X given by the system:

$$z_1 - z_0 x_j(\sigma,\tau) = 0$$

$$z_2 - z_0 y_j(\sigma,\tau) = 0.$$

Denote by $P^{(j)}$ the closure in Y of $\tilde{s}^{-1}(\tilde{P}^{(j)} - \mathfrak{S})$. $P^{(j)}$ is given in Y_U by the following system of equations:

$$y' = \prod_{\ell=1}^{j} (\sigma - \rho_1^{\ell}\tau) e_j'(\sigma,\tau),$$

$$x' = \prod_{\ell=j+1}^{n_1} (\sigma - \rho_1^{\ell}\tau) e_j''(\sigma,\tau),$$

$$\xi_{11} = \prod_{\ell=2}^{j} (\sigma - \rho_1^{\ell}\tau) e_j''^{-1} \xi_{01}, \qquad \xi_{0j+1} = \prod_{\ell=j+1}^{j+1} (\sigma - \rho_1^{\ell}\tau) e_j'' \xi_{1j+1}$$

- -

$$\xi_{1j-1} = \prod_{\ell=j}^{j} (\sigma - \rho_1^{\ell}\tau) e_j''^{-1} \xi_{0j-1}, \quad \xi_{0n_1-1} = \prod_{\ell=j+1}^{n_1} (\sigma - \rho_1^{\ell}\tau) e_j'' \xi_{1n_1-1}$$

$$\xi_{1j} = e_j''^{-1} \xi_{0j} \ ,$$

It is clear that $P^{(j)}$ is a cross-section of $\mathfrak{P}': Y \longrightarrow M_\epsilon$ and that $P^{(j)} \cap \mathfrak{P}'^{-1}(0,0) \in \tilde{C}_0^{(j)}$.

Let $\tilde{P}^{(0)} = \{a \in X \mid z_0(a) = 0\}$, $P^{(0)} = \tilde{s}^{-1}(\tilde{P}^{(0)})$ and $P^{(i)}(\sigma,\tau) = P^{(i)} \cap \mathfrak{P}'^{-1}(\sigma,\tau)$, $(\sigma,\tau) \in M_\epsilon$,

$$Y^{(i)} = \tilde{s}^{-1}(X_{(0)}) \cup \tilde{C}_0^{(i)}, \quad i = 0,1,\cdots,n_1-1.$$

It is clear that $\{Y^{(0)},\cdots,Y^{(n_1-1)}\}$ form an open covering of $Y_{(0)} = \tilde{s}^{-1}(X_{(0)}) \cup \tilde{C}_0 = \bigcup_{(\sigma,\tau)\in M_\epsilon} \tilde{C}_{(\sigma,\tau)}$.

Let $M_1 = M_\epsilon \times \mathbb{C}^*$, $S_1 = \Sigma_\epsilon \times \mathbb{C}^*$, $\mathcal{D}$ be a cyclic group of analytic automorphisms of $M_1' = M_1 - S_1$ generated by the transformatio

$(\sigma,\tau,w) \longrightarrow (\sigma,\tau,(\sigma^{n_1}-\tau^{n_1})w)$. It is easy to verify that the Kodaira condition for existence of Kodaira factor-space $M_1/\mathcal{D}$ is satisfied here (we gave references and little more explanation on page 126).

Using $e^{2\pi i\omega(\sigma,\tau)} = \sigma^{n_1}-\tau^{n_1}$, we define for $i = 0,\cdots,n_1-1$ holomorphic maps $\Psi^{(i)}: Y^{(i)} \longrightarrow M_1/\mathcal{D}$ by

$$\Psi^{(i)}(a) = [(\sigma(a),\tau(a);\ \exp(2\pi i \int_{(i)(\sigma(a),\tau(a)}^{a} \alpha_1(\sigma(a),\tau(a)))]$$

where $\int_{P^{(i)}(\sigma(a),\tau(a)}$ is taken along a path in $\tilde{C}_{(\sigma(a),\tau(a))} \cap Y^{(i)}$.

It is clear that each $\Psi^{(i)}$ is onto and 1-1 map. Hence $\Psi^{(i)}$ is an isomorphism of complex manifolds.

Denote $w^{(j)}(a) = \exp(\int_{P^{(j)}(\sigma(a),\tau(a)}^{a} \alpha_1(\sigma(a),\tau(a)))$, $a \in Y^{(j)}$,

$\int_{P^{(j)}(\sigma(a),\tau(a))}^{a}$ is taken along a path in $\tilde{C}_{(\sigma(a),\tau(a))} \cap Y^{(j)}$,

$j = 0,\cdots,n_1-1$.

Let $\rho_2 = \rho_1^{b}$. We define analytic isomorphisms:

$$h_{1j}: Y^{(j)} \longrightarrow Y^{(j+b)} (\subset Y)$$

where the index j is considered as an element of the cyclic group $\mathbb{Z}_{n_1} = \mathbb{Z}/(n_1)$ of order n_1, that is, $Y^{(n_1)} = Y^{(0)}$, $Y^{(n_1+1)} = Y^{(1)}$, etc., by

$$\text{(2)}\qquad \begin{aligned} \sigma(h_{1j}(a)) &= \rho_2\sigma(a) \\ \tau(h_{1j}(a)) &= \tau(a) \\ w^{(j+b)}(h_{1j}(a)) &= w^{(j)}(a)\rho_2^j\tilde{c}, \end{aligned}$$

where

$$\tilde{c} = \exp\frac{\pi i k' b\,(m-1)}{m}$$

Let $a \in Y^{(j)} \cap Y^{(j')}$, $j \neq j'$ (that is, $(\sigma(a),\tau(a) \neq (0,0))$. We shall prove that $h_{1j}(a) = h_{1j'}(a)$. Let $a_j = h_{1j}(a)$, $a_{j'} = h_{1j'}(a)$. We have

$$\frac{w^{(j+b)}(a_j)}{w^{(j'+b)}(a_{j'})} = \exp\Bigg(2\pi i\Bigg(\int_{p^{(j+b)}(\rho_2\sigma(a),\tau(a))}^{a_j} \alpha_1(\rho_2\sigma(a),\tau(a)) - \int_{p^{(j'+b)}(\rho_2\sigma(a),\tau(a))}^{a_{j'}} \alpha_1(\rho_2\sigma(a),\tau(a))$$

$$= \exp(2\pi i \int_{a_{j'}}^{a_j} \alpha_1(\rho_2\sigma(a),\tau(a)) \cdot \exp(2\pi i \int_{p^{(0)}(\rho_2\sigma(a),\tau(a))}^{p^{(j'+b)}(\rho_2(\sigma(a)),\tau(a))} \alpha_1(\rho_2\sigma(a),\tau(a))$$

$$- \int_{p^{(0)}(\rho_2\sigma(a),\tau(a))}^{p^{(j+b)}(\rho_2(\sigma(a),\tau(a))} \alpha_1(\rho_2\sigma(a),\tau(a)) =$$

$$= \exp(2\pi i \int_{a_{j'}}^{a_j} \alpha_1(\rho_2(\sigma(a),\tau(a)) \cdot \frac{w_{j'+b}(\rho_2\sigma(a),\tau(a))}{w_{j+b}(\rho_2\sigma(a),\tau(a))} \quad ,$$

where we assume $w_0 = w_{n_1} = 1$.

From (2) we get

$$\frac{w^{(j+b)}(a_j)}{w^{(j'+b)}(a'_j)} = \rho_2^{j-j'} \frac{w^{(j)}(a)}{w^{(j')}(a)} = \rho_2^{j-j'} \frac{w_{j'}(\sigma(a),\tau(a))}{w_j(\sigma(a),\tau(a))} .$$

Without loss of generality we can assume that $(\sigma(a))^{n_1}-(\tau(a))^{n_1}\neq 0$ (that is, we have to prove only that the maps h_{1j} and $h_{1j'}$ coincide on an open dense subset of $Y^{(j)} \cap Y^{(j')}$). We have

$$w_{j+b}(\rho_2\sigma(a),\tau(a)) \equiv \prod_{\ell=1}^{j+b}(\rho_2\sigma(a)-\rho_1^{\ell}\tau(a))^{-1}(\mathrm{mod}((\sigma(a))^{n_1}-(\tau(a))^{n_1}))$$

(where congruence is considered in the group $\mathbb{C}^*$),

$$w_j(\sigma(a),\tau(a)) \equiv \prod_{\ell=1}^{j}(\sigma(a)-\rho_1^{\ell}\tau(a))^{-1}(\mathrm{mod}((\sigma(a))^{n_1}-(\tau(a))^{n_1})).$$

Because

$$\prod_{\ell=1}^{j+b}(\rho_2\sigma-\rho_1^{\ell}\tau) = \rho_2^{j+b} \prod_{\ell=1}^{j+b}(\sigma-\rho_1^{\ell-b}\tau) =$$

$$= \rho_2^{j+b} \prod_{\ell=1-b}^{0}(\sigma-\rho_1^{\ell}\tau)\cdot \prod_{\ell=1}^{j}(\sigma-\rho_1^{\ell}\tau)$$

we obtain that

$$\frac{w_{j+b}(\rho_2\sigma(a),\tau(a))}{w_j(\sigma(a),\tau(a))} \equiv \rho_2^{-j-b} \prod_{\ell=1-b}^{0} (\sigma_{(a)}-\rho_1^{\ell}\tau(a))(\mathrm{mod}((\sigma(a))^{n_1}-(\tau(a))^{n_1}))$$

and (by the same reasons)

$$\frac{w_{j'+b}(\rho_2\sigma(a),\tau(a))}{w_{j'}(\sigma(a),\tau(a))} \equiv \rho_2^{-j'-b} \prod_{\ell=1-b}^{0} (\sigma(a)-\rho_1^{\ell}\tau(a))(\mathrm{mod}((\sigma(a))^{n_1}-(\tau(a))^{n_1}).$$

Hence

$$\frac{w_{j'+b}(\rho_2\sigma(a),\tau(a))}{w_{j+b}(\rho_2\sigma(a),\tau(a))} \equiv \rho_2^{j-j'} \frac{w_{j'}(\sigma(a),\tau(a))}{w_j(\sigma(a),\tau(a))}(\mathrm{mod}((\sigma(a))^{n_1}-(\tau(a))^{n_1})$$

and

$$\exp(2\pi i \int_{a_{j'}}^{a_j} \alpha_1(\rho_2(\sigma(a),\tau(a))) \equiv 1(\mathrm{mod}((\sigma(a))^{n_1}-(\tau(a))^{n_1}).$$

We see that $a_j = a_j'$, that is, $h_{1j}(a) = h_{1j'}(a)$.

Now we can define an analytic isomorphism $h_1': Y_{(o)} \longrightarrow Y_{(o)}$ by $h_1'(a) = h_{1j}(a)$ where $a \in Y^{(j)}$.

For $(\sigma,\tau) \neq (0,0)$ let $q'(\sigma,\tau)$ be the singular point of $C_{(\sigma,\tau)}$. We extend h_1' to $h_1'': Y_{(o)}' \longrightarrow Y_{(o)}'$ where $Y_{(o)}' = Y_{(o)} \cup (\bigcup_{(\sigma,\tau)\in M_\epsilon^*} q'(\sigma,\tau))$ by $h_1''(q'(\sigma,\tau)) = q'(\rho_2\sigma,\tau)$, $h_1''|_{Y_{(o)}} = h_1'$. In order to prove that h_1'' is analytic, it is enough to prove that h_1'' is continuous. Let $\{a^{(r)}, r = 1,2,\cdots\}$, be an infinite sequence with $a^{(r)} \in Y_{(o)}$, $\lim_{r\to\infty} a^{(r)} = q'(\sigma,\tau)$. We have to prove that $\lim_{r\to\infty} h_1'(a^{(r)}) = q'(\rho_2\sigma,\tau)$. Suppose that it is not true. Then there exists a subsequence $\{a^{(r_p)}, p = 1,2,\cdots,\}$ such that $\lim_{p\to\infty} h_1'(a^{(r_p)}) = a' \in \tilde{C}_{(\rho_2\sigma,\tau)}$. We obtain $\lim_{p\to\infty} a^{(r_p)} = h_1'^{-1}(a') \in \tilde{C}_{(\sigma,\tau)}$. Contradiction.

Now we extend h_1'' to $h_1 : Y \longrightarrow Y$ by $h_1(q_o^{(j)}) = q_o^{(j+b)}$, $j = 0,1,\cdots,n_1-1$, ($q_o^{(n_1)} = q_o^{(o)}$, $q_o^{(n_1+1)} = q_o^{(1)}$, etc.), $h_1|_{Y'_{(o)}} = h_1''$.

As before, we have to prove that h_1 is continuous. Suppose that it is not true at some $q_o^{(j)}$, $j \in (0,1,\cdots,n_1-1)$. Take an infinite sequence $\{U_r, r = 1,2,\cdots\}$ where each U_r is a relatively compact open neighborhood of $q_o^{(j)}$ in Y,

$$\bigcap_{r=1}^{\infty} \overline{U}_r = q_o^{(j)}, \quad U_{r+1} \subset U_r, \quad U_r \cap \left(\bigcup_{\ell=0}^{n_1-1} q_o^{(\ell)}\right) = q_o^{(j)}$$

and $U_r - q_o^{(j)}$ is connected for any $r = 1,2,\cdots$.

Denote $U_r' = h_1''(U_r - q_o^{(j)})$. It is easy to see that the closure $\overline{U}_r'$ of U_r' is compact. A direct calculation shows that if $\{a_r, r = 1,2,\cdots\}$ is an infinite sequence with $a_r \in \tilde{C}_o^{(j)}$, $\lim_{r\to\infty} a_r = q_o^{(j)}$ then $\lim_{r\to\infty} h_1'(a_r) = q_o^{(j+b)}$.

Hence

$$q_o^{(j+b)} \in \bigcap_{r=1}^{\infty} \overline{U}_r'.$$

Suppose there is a point $a' \in \bigcap_{r=1}^{\infty} \overline{U}_r'$ such that $a' \notin \bigcup_{\ell=0}^{n_1-1} q_o^{(\ell)}$, that is,

$$h_1''^{-1}(a') \notin \bigcup_{\ell=0}^{n_1-1} q_o^{(\ell)}.$$

Then there exists a positive integer r_o such that $\overline{U}_{r_o} \not\ni h_1''^{-1}(a')$. Take a sequence $\{a_r', r = 1,2,\cdots\}$ with $a_r' \in U_{r_o}'$, $\lim_{r\to\infty} a_r' = a'$. We have

$h_1''^{-1}(a') = \lim_{r\to\infty} h_1''^{-1}(a'_r) \in \overline{U}_{r_o}$ (because $h_1''^{-1}(a'_r) \subset h_1''^{-1}(U'_{r_o}) = U_{r_o} - q_o^{(j)}$.

Contradiction. Hence

$$\bigcap_{r=1}^{\infty} \overline{U}'_r \subset \bigcup_{\ell=0}^{n_1-1} q_o^{(\ell)}.$$

But each U'_r is connected. Thus each $\overline{U}'_r$ is connected and from $\ldots \supset \overline{U}'_r \supset \overline{U}'_{\ell+1} \supset \ldots$ and the compactness of $\overline{U}'_1$ we get that $\bigcap_{r=1}^{\infty} \overline{U}'_r$ is connected. We see that $\bigcap_{r=1}^{\infty} \overline{U}'_r = q_o^{(j+b)}$. It follows from this that h_1 is continuous at $q_o^{(j)}$. Thus we proved that $h_1 \colon Y \longrightarrow Y$ is an analytic automorphism of Y.

Since Y is homotopically equivalent to $\mathfrak{F}'^{-1}(0,0)$ we have that $\pi_1(Y)$ is isomorphic to $\mathbb{Z}$. Let $\beta\colon Z \longrightarrow Y$ be an unramified covering of Y corresponding to the subgroup $d\mathbb{Z}$ of $\mathbb{Z}$,
$\mathfrak{F}'' = \mathfrak{F}' \cdot \beta$, $Z(\tau) = \{a \in Z \mid \tau(\mathfrak{F}''(a)) = \tau\}$

$$\mathfrak{F}''_\tau = \mathfrak{F}''|_{Z(\tau)} \colon Z(\tau) \longrightarrow D_{\epsilon,\tau}.$$

It is easy to see that all fibers of $\mathfrak{F}''$ are connected and that τ is a holomorphic function on Z without critical values. Hence each $Z(\tau)$ ($|\tau| \le \epsilon$) is nonsingular, for any $a \in D_{\epsilon,\tau} - D_{\epsilon,\tau} \cap \Sigma_\epsilon$ the fiber $\mathfrak{F}''^{-1}(a)$ is a nonsingular elliptic curve (and a regular fiber of $\mathfrak{F}''_\tau$), $\mathfrak{F}''^{-1}_\tau((\rho_1^\ell\tau,\tau))$, $\tau \ne 0$, $\ell = 0,1,\cdots,n_1-1$, are singular fibers of $\mathfrak{F}''_\tau$ which have type I and $\mathfrak{F}''^{-1}((0,0))$ is a singular fiber of

$\mathfrak{F}_o'': Z(0) \longrightarrow D_{\epsilon,0}$ which is of type $I_{n_1 d} = I_{mb_1 d} = I_{mb} = I_n$.

We can assume that $q_o^{(0)},\cdots,q_o^{(n_1-1)}$ are contained in some circle c_o of $C_{(0,0)}$, which is a generator of $\pi_1(Y)$, and that the order of $q_o^{(0)},\cdots,q_o^{(n-1)}$ in c_o coincides with the order of their indexes.

Denote $c_o' = \beta^{-1}(c_o)$ and let $\beta^{-1}(q_o^{(j)}) = \bigcup_{r=0}^{d-1} q(rn_1+j)$.

We can assume that the points

$$q(0),q(1),\cdots,q(n_1-1),q(n_1),\cdots,q(2n_1-1),\cdots,q((d-1)n_1),\cdots,q(dn_1-1)$$

have the same order on c_o' as their indexes. Because $\pi_1(Y) \approx \mathbb{Z}$ has only one subgroup of index n_1 we obtain that there exists unique analytic automorphism $h: Z \longrightarrow Z$ such that the diagram

$$\begin{array}{ccc} Z & \xrightarrow{\ h\ } & Z \\ \downarrow{\scriptstyle\beta} & & \downarrow{\scriptstyle\beta} \\ Y & \xrightarrow[h_1]{} & Y \end{array}$$

is commutative and $h(q(0)) = q(b)$.

Let G_1 (corresp. G) be a cyclic group of analytic automorphisms of Y (corresp. Z) generated by h_1 (corresp. h). A direct verification shows that G_1 is isomorphic to $\mathbb{Z}/m\mathbb{Z}$ and that any G_1-orbit has exactly m points. We get that G is isomorphic to $\mathbb{Z}/m\mathbb{Z}$ and that any G-orbit has exactly m points.

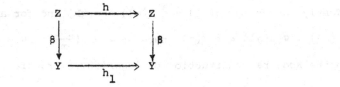

Let $\widetilde{M}_\epsilon = \{(\tilde{\sigma},\tilde{\tau}) \in \mathbb{C}^2 \,|\, |\tilde{\sigma}| < \epsilon^m, \; |\tilde{\tau}| < \epsilon\}$, $\tilde{\pi}: M_\epsilon \longrightarrow \widetilde{M}_\epsilon$ be given by $\tilde{\pi}(\sigma,\tau) = (\sigma^m,\tau)$, $\tilde{Z} = Z/G$, $\tilde{\beta}: Z \longrightarrow Z/G$ be the canonical map.

It is easy to see that we have a commutative diagram of holomorphic maps

$$\begin{array}{ccc} Z & \xrightarrow{\tilde{\beta}} & \tilde{Z} \\ {\scriptstyle \mathfrak{F}''}\big\downarrow & & \big\downarrow{\scriptstyle \tilde{\mathfrak{F}}} \\ M_\epsilon & \xrightarrow[\tilde{\pi}]{} & \widetilde{M}_\epsilon \end{array} .$$

Let

$$\tilde{Z}(\tau) = \{a \in Z \,|\, \tilde{\tau}(\tilde{\mathfrak{F}}(a)) = \tau\}, \quad \tilde{\mathfrak{F}}(\tau) = \tilde{\mathfrak{F}}\big|_{\tilde{Z}(\tau)}: \tilde{Z}(\tau) \longrightarrow D_{\epsilon,\tau},$$

$g = h^k$, $g_o = g\big|_{Z(o)}: Z(o) \longrightarrow Z(o)$. We can assume $0 < k < m$. Evidently we have $g_o(q(o)) = g(q(o)) = q(kb)$ and for any $a \in Z(o)$ $\sigma(g_o(a)) = \sigma(g(a)) = \rho \cdot \sigma(a)$ where $\rho = \exp(\frac{2\pi i}{m})$. Now comparing with the Kodaira construction of logarithmic transform and using Kodaira's arguments in ([14], p. 769) we obtain that $\tilde{\mathfrak{F}}(o): \tilde{Z}(o) \longrightarrow D_{\epsilon,o}$ has unique non-regular fiber at $\sigma = 0$ and this fiber is of type ${}_mI_b$ with the desired invariant k. We see that we can identify $f': V' \longrightarrow D_\epsilon$ with $\tilde{\mathfrak{F}}(o): \tilde{Z}(o) \longrightarrow D_{\epsilon,o}$.

If $\tau \neq 0$ the only multiple fiber of $\tilde{\mathfrak{F}}(\tau): \tilde{Z}(\tau) \longrightarrow D_{\epsilon,\tau}$ corresponds to $\sigma = 0$. Hence it is non-singular.

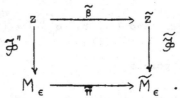

Take positive integers $\epsilon_i < \epsilon$, $i = 1,2,3$, such that $\epsilon_1 > \epsilon_2$. Let $A = \{(\sigma,\tau) \in \mathcal{M}_\epsilon | \epsilon_2 < |\sigma| < \epsilon_1, |\tau| < \epsilon_3\}$, $P_A^{(o)} = P^{(o)}|_A$,

$Y_A' = \mathfrak{F}'^{-1}(A) - P_A^{(o)}$. Y_A' is given in $A \times \mathbb{C}^2$ by the equation

$$y^2 = 4x^3 - 3x - B(\sigma,\tau).$$

Because A is a Stein manifold we have that $A \times \mathbb{C}^2$ and $Y_A' \subset A \times \mathbb{C}^2$ are Stein manifolds. It follows from this that $\beta^{-1}(Y_A')$ is a Stein manifold.

Let $Z_A' = \bigcap_{j=0}^{m-1} h^j(\beta^{-1}(Y_A'))$. Z_A' is also a Stein manifold, because it is isomorphic to a closed analytic subvariety in $\prod_{j=0}^{m-1} h^j(\beta^{-1}(Y_A'))$. It is clear that $h(Z_A') = Z_A'$, $\mathfrak{F}''(Z_A') = A$ and $\mathfrak{F}''^{-1}(A) - Z_A'$ is a proper analytic subvariety of $\mathfrak{F}''^{-1}(A)$.

We obtain that $\tilde{Z}_A' = \tilde{\beta}(Z_A')$ (which is isomorphic to Z_A'/G) is a Stein manifold and $\tilde{\beta}(\mathfrak{F}''^{-1}(A)) - \tilde{Z}_A'$ is a proper analytic subset in $\tilde{\beta}(\mathfrak{F}''^{-1}(A))$.

Let $\tilde{A} = \{(\tilde{\sigma},\tilde{\tau}) \in \tilde{\mathcal{M}}_\epsilon | \epsilon_2^m < |\tilde{\sigma}| < \epsilon_1^m, |\tilde{\tau}| < \epsilon_3\}$,

$$\tilde{A}_o = \{(\tilde{\sigma},\tilde{\tau}) \in \tilde{A} | \tilde{\tau} = o\}, \quad \tilde{Z}_{\tilde{A}} = \tilde{\mathfrak{F}}^{-1}(\tilde{A}), \quad \tilde{Z}_{\tilde{A},o} = \tilde{\mathfrak{F}}^{-1}(\tilde{A}_o),$$

$$\tilde{\mathfrak{F}}_{\tilde{A}} = \tilde{\mathfrak{F}}|_{\tilde{Z}_{\tilde{A}}} : \tilde{Z}_{\tilde{A}} \longrightarrow \tilde{A}, \quad \tilde{\mathfrak{F}}_{\tilde{A},o} = \tilde{\mathfrak{F}}|_{\tilde{Z}_{\tilde{A},o}} : \tilde{Z}_{\tilde{A},o} \longrightarrow \tilde{A}_o .$$

From the Kodaira construction of logarithmic transform ([14], p. 770) it follows that there exists a cross-section η_o of $\tilde{\mathfrak{F}}_{\tilde{A},o}: \tilde{Z}_{\tilde{A},o} \longrightarrow \tilde{A}_o$ such that η_o is not contained in $\tilde{Z}_{\tilde{A}} - \tilde{Z}_A'$. Since $\dim_{\mathbb{C}} \eta_o = 1$ we have that

$\dim_{\mathbb{C}} \eta_o \cap (\tilde{Z}_{\tilde{A}} - \tilde{Z}'_A) = 0$. Thus (changing if necessary ϵ_i, $i = 1,2,3$) we can assume that $\eta_o \subset \tilde{Z}'_A$. Let $\tilde{Z}'_{A,o} = \tilde{Z}'_A \cap \tilde{Z}_{\tilde{A},o}$. Taking ϵ_3 smaller we can prove that $\tilde{Z}'_A$ is homeomorphic to $\tilde{Z}'_{A,o} \times D_{\epsilon_3}$ $(D_{\epsilon_3} = \{\tilde{\tau} \in \mathbb{C} \,|\, |\tilde{\tau}| < \epsilon_3\}$ and $\tilde{Z}'_A \longrightarrow D_{\epsilon_3}$ given by $x \longrightarrow \tilde{\tau}\,\tilde{\mathfrak{F}}(x)$ coincides with the projection $\tilde{Z}'_{A,o} \times D_{\epsilon_3} \longrightarrow D_{\epsilon_3}$. We see that the canonical map $H^2(\tilde{Z}'_A, \mathbb{Z}) \longrightarrow H^2(\tilde{Z}'_{A,o}, \mathbb{Z})$ (corresponding to the embedding $\tilde{Z}'_{A,o} \subset \tilde{Z}'_A$) is an isomorphism. Since $\tilde{Z}'_{A,o}$ and $\tilde{Z}'_A$ are Stein manifolds we get that the canonical map $\mathrm{Pic}(\tilde{Z}'_A) \longrightarrow \mathrm{Pic}(\tilde{Z}'_{A,o})$ is an isomorphism and there exists a complex line bundle $[\mathcal{E}]$ on $\tilde{Z}'_A$ such that $[\mathcal{E}]\big|_{Z'_{A,o}} = [\eta_o]$ where $[\eta_o]$ is a line bundle on $\tilde{Z}'_{A,o}$ corresponding to the curve $\eta_o \subset \tilde{Z}'_{A,o}$ $(\dim_{\mathbb{C}} Z'_{A,o} = 2)$. Let φ_o be a cross-section of $[\eta_o]$ with zero-locus η_o. From the exact sequence

$$H^o(\tilde{Z}'_A, \mathcal{O}[\mathcal{E}]) \longrightarrow H^o(\tilde{Z}'_{A,o}, \mathcal{O}[\mathcal{E}]\big|_{\tilde{Z}'_{A,o}}) \longrightarrow 0$$

we have that there exists a cross-section $\varphi \in H^o(\tilde{Z}'_A, \mathcal{O}[\mathcal{E}])$ such that $\varphi\big|_{\tilde{Z}'_{A,o}} = \varphi_o$. Let $\eta = \{x \in \tilde{Z}'_A \,|\, \varphi(x) = 0\}$. It is clear that

$$\eta \cap \tilde{Z}'_{A,o} = \eta_o \quad \text{and} \quad \dim_{\mathbb{C}} \eta = 2.$$

Changing if necessary ϵ_i, $i = 1,2,3$, we obtain that η is closed in $\tilde{Z}_{\tilde{A}}$ and $\tilde{\mathfrak{F}}(\eta) = \tilde{A}$. Because $(\eta . \tilde{\mathfrak{F}}^{-1}(\tilde{\sigma},\tilde{\tau})) = 1$, where $(\tilde{\sigma},\tilde{\tau}) \in \tilde{A}$, we have that η is a cross-section of $\tilde{\mathfrak{F}}_{\tilde{A}}: \tilde{Z}_{\tilde{A}} \longrightarrow \tilde{A}$.

In order to finish the proof of Lemma 4 we shall use the following

<u>Lemma 5</u>. Let $A = \{(\sigma,\tau) \in \mathbb{C}^2 \,|\, \epsilon_2 < |\sigma| < \epsilon_1,\ |\tau| < \epsilon_3\}$, $\epsilon_i > 0$, $i = 1,2,3$, and $\mathfrak{P}: Z \longrightarrow A$ be a proper surjective holomorphic map of complex manifolds such that $(d\mathfrak{P})_x$ is an epimorphism for any $x \in Z$ and all fibers of $\mathfrak{P}$ are elliptic curves. Let $\mathcal{J}(\sigma,\tau)$ be a holomorphic function in $\overline{A}$, which for any $(\sigma,\tau) \in A$ is equal to the absolute invariant of $\mathfrak{P}^{-1}(\sigma,\tau)$. Suppose that $\mathcal{J}(\sigma,\tau)\big|_{\tau=0}$ is not a constant (as function of σ) and that there exists a holomorphic cross-section η of $\mathfrak{P}: Z \longrightarrow A$. Then there exist positive numbers ϵ_i', $i = 1,2,3$, such that $\epsilon_3' \leq \epsilon_3$, $\epsilon_2 \leq \epsilon_2' < \epsilon_1' \leq \epsilon_1$, and if $A_o' = \{(\sigma,\tau) \in A \,|\, \tau = 0,\ \epsilon_2' < |\sigma| < \epsilon_1'\}$, $Z_o' = \mathfrak{P}^{-1}(A_o')$, $\mathfrak{P}_o' = \mathfrak{P}|_{Z_o'}: Z_o' \longrightarrow A_o'$, $D_{\epsilon_3'} = \{\tau \in \mathbb{C} \,|\, |\tau| < \epsilon_3'\}$, then there exists a commutative diagram of holomorphic maps

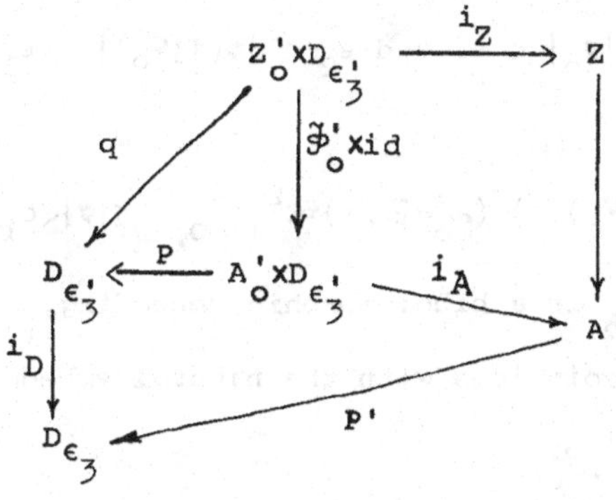

where i_Z and i_A are biholomorphic embeddings,

$D_{\epsilon_3} = \{\tau \in \mathbb{C} \,|\, |\tau| < \epsilon_3\}$, $i_D(\tau) = \tau$, $p'(\sigma,\tau) = \tau$, p,q are canonical projections, $i_A|_{A_o' \times 0}: A_o' \times 0 \longrightarrow p'^{-1}(0)$ coincides with the natural embedding $A_o' \longrightarrow p'^{-1}(0)$ and $i_Z|_{Z_o' \times 0}: Z_o' \times 0 \longrightarrow \mathfrak{F}^{-1}(p'^{-1}(0))$ coincides with the natural embedding $Z_o' \longrightarrow \mathfrak{F}^{-1}(p'^{-1}(0))$.

Proof of Lemma 5.

Changing, if necessary, ϵ_i, $i = 1,2,3$, we may assume that $\frac{\partial \mathcal{J}}{\partial \sigma}$ has no zeroes in A. There exist positive numbers ϵ_i', $i = 1,2,3$, $\epsilon_3' < \epsilon_3$, $\epsilon_2 < \epsilon_2' < \epsilon_1' < \epsilon_1$ such that the differential equation

$$\frac{d\varphi}{d\tau} = - \frac{\frac{\partial \mathcal{J}}{\partial \tau}(\varphi,\tau)}{\frac{\partial \mathcal{J}}{\partial \sigma}(\varphi,\tau)}$$

has unique holomorphic solution $\varphi = \varphi(\tau;\sigma_o)$ with $\varphi(0;\sigma_o) = \sigma_o$, $|\tau| < \epsilon_3'$, $\epsilon_2' < |\sigma_o| < \epsilon_1'$ and $\epsilon_2 < |\varphi(\tau;\sigma_o)| < \epsilon_1$ for $|\tau| < \epsilon_3'$, $\epsilon_2' < |\sigma_o| < \epsilon_1'$.

Define $i_A: A_o' \times D_{\epsilon_3'} \to A$ by $i_A(\sigma_o,\tau) = (\varphi(\tau;\sigma_o),\tau)$ $(A_o' = \{(\sigma,\tau) \in \mathbb{C}^2 \,|\, \tau = 0,\ \epsilon_2' < |\sigma| < \epsilon_1\}$.

It is easy to see that i_A is a biholomorphic embedding, $i_A|_{A_o' \times 0}: A_o' \times 0 \longrightarrow p'^{-1}(0)$ coincides with the natural embedding $A_o' \longrightarrow p'^{-1}(0)$ and $p' i_A = i_D p$.

Let $Z' = Z \times_A (A_o' \times D_{\epsilon_3'})$ (corresponding to $\mathcal{F}$ and i_A), $i': Z' \longrightarrow Z$, $\mathcal{F}': Z' \longrightarrow A_o' \times D_{\epsilon_3}'$ be canonical projections, $\mathcal{J}' = i_A^* \mathcal{J}$, $\eta' = (i')^*\eta$.

We have

$$\frac{\partial}{\partial\tau}\mathcal{J}(\varphi(\tau;\sigma_o);\tau) = \frac{\partial\mathcal{J}}{\partial\sigma}\frac{\partial\varphi}{\partial\tau} + \frac{\partial\mathcal{J}}{\partial\tau} = \frac{\partial\mathcal{J}}{\partial\sigma}\left(-\frac{\frac{\partial\mathcal{J}}{\partial\tau}}{\frac{\partial\mathcal{J}}{\partial\sigma}}\right) + \frac{\partial\mathcal{J}}{\partial\tau} = 0.$$

Hence $\mathcal{J}'(\sigma_o,\tau)$ does not depend on τ. Let $p'': A_o' \times D_{\epsilon_3'} \longrightarrow A_o'$ be a canonical projection, $\mathcal{J}'' = \mathcal{J}'|_{A_o' \times 0}$. We see that $\mathcal{J}' = (p'')^* \mathcal{J}''$.

Changing, if necessary, ϵ_i', $i = 1,2,3$, we can assume that there exists an open covering $\{U_i, i \in \mathfrak{I}\}$ of A_o' such that if $V_i = U_i \times D_{\epsilon_3'}$, $i \in \mathfrak{I}$, then there exists a continuous family $\{\delta_{1i}(\sigma,\tau), \delta_{2i}(\sigma,\tau), (\sigma,\tau) \in V_i\}$, where $\delta_{1i}(\sigma,\tau), \delta_{2i}(\sigma,\tau) \in H_1(\mathcal{F}'^{-1}(\sigma,\tau,\mathbb{Z})$, $\delta_{1i}(\sigma,\tau), \delta_2(\sigma,\tau)$ generate $H_1(\mathcal{F}'^{-1}(\sigma,\tau),\mathbb{Z})$, and a holomorphic family $\{\alpha_i(\sigma,\tau), (\sigma,\tau) \in V_i\}$ where $\alpha_i(\sigma,\tau)$ is a non-zero holomorphic differential on $\mathcal{F}'^{-1}(\sigma,\tau)$. We can assume also that if $(\sigma,\tau) \in V_i \cap V_j$, $i,j \in \mathfrak{I}$, then

$$\delta_{1j}(\sigma,\tau) = a_{ij}\delta_{1i}(\sigma,\tau) + b_{ij}\delta_{2i}(\sigma,\tau),$$

$$\delta_{2j}(\sigma,\tau) = c_{ij}\delta_{1i}(\sigma,\tau) + d_{ij}\delta_{2i}(\sigma,\tau)$$

where $a_{ij}, b_{ij}, c_{ij}, d_{ij} \in \mathbb{Z}$ depend only of i and j, and that

$$\operatorname{Im} \frac{\int_{\delta_{1i}(\sigma,\tau)} \alpha_i(\sigma,\tau)}{\int_{\delta_{2i}(\sigma,\tau)} \alpha_i(\sigma,\tau)} > 0 , \quad \forall i \in J.$$

Let $\omega_{1i}(\sigma,\tau) = \int_{\delta_{1i}(\sigma,\tau)} \alpha_i(\sigma,\tau)$, $\omega_{2i}(\sigma,\tau) = \int_{\delta_{2i}(\sigma,\tau)} \alpha_i(\sigma,\tau)$,

$$\omega_i(\sigma,\tau) = \frac{\omega_{1i}(\sigma,\tau)}{\omega_{2i}(\sigma,\tau)}, \quad \tilde{\alpha}_i(\sigma,\tau) = \frac{1}{\omega_{2i}(\sigma,\tau)}\alpha_i(\sigma,\tau).$$

Note that $\omega_i(\sigma,\tau)$ does not depend on τ. If it would not be so, we would get that for some $\sigma_o \in A_o'$, $\omega_i(\sigma_o,\tau)$ is a non-constant holomorphic function, that is, it has a continuum set of values. Then $\mathcal{J}'|_{\sigma=\sigma_o}$ is not a constant. Contradiction (with $\frac{\partial \mathcal{J}'}{\partial \tau} = 0$). We can write $\omega_i(\sigma,\tau) = \omega_i(\sigma)$.

Let $(\sigma,\tau) \in V_i \cap V_j$, $i,j \in J$. We have

$$\int_{\delta_{2j}(\sigma,\tau)} \tilde{\alpha}_i(\sigma,\tau) = c_{ij}\omega_i(\sigma) + d_{ij}.$$

This shows that $c_{ij}\omega_i(\sigma)+d_{ij} \neq 0$. Since

$$\int_{\delta_{2j}(\sigma,\tau)} (\tilde{\alpha}_j(\sigma,\tau) - \frac{1}{c_{ij}\omega_i(\sigma)+d_{ij}}\, \tilde{\alpha}_i(\sigma,\tau)) = 0$$

we obtain that $\tilde{\alpha}_j(\sigma,\tau) = \frac{1}{c_{ij}\omega_i(\sigma)+d_{ij}}\, \tilde{\alpha}_i(\sigma,\tau)$.

Let G_i be a group of analytic automorphisms of $U_i \times \mathbb{C}$ which have the following form:

$$(\sigma,\zeta) \longrightarrow (\sigma,\ \zeta + n_1\omega_i(\sigma) + n_2),\quad n_1, n_2 \in \mathbb{Z}\ .$$

Define a holomorphic map:

$$\Psi_i\colon\ \mathfrak{F}'^{-1}(V_i) \longrightarrow U_i \times \mathbb{C}/G_i$$

by $$\Psi_i(x) = [\sigma(\mathfrak{F}'(x)),\ \int_{\eta'(\mathfrak{F}'(x))}^{x} \tilde{\alpha}_i(\mathfrak{F}'(x))]$$

where $\int_{\eta'(\mathfrak{F}'(x))}^{x}$ is taken along a path in $\mathfrak{F}'^{-1}(\mathfrak{F}'(x))$.

Let $\Psi_{oi} = \Psi_i|_{\mathfrak{F}'^{-1}(U_i \times 0)}\colon \mathfrak{F}'^{-1}(U_i \times 0) \longrightarrow U_i \times \mathbb{C}/G_i$.

It is clear that Ψ_{oi} is an isomorphism. Denote $T_i = \Psi_{oi}^{-1}$ and define $\nu_i\colon\ \mathfrak{F}'^{-1}(V_i) \longrightarrow Z_o' \times D_{\epsilon_3'}$ by $\nu_i(x) = (T_i\Psi_i(x), \tau(\mathfrak{F}'(x)))$.

Let $x \in \mathfrak{F}'^{-1}(V_i) \cap \mathfrak{F}'^{-1}(V_j)$, $i,j \in J$. We shall prove that $\nu_i(x) \equiv \nu_j(x)$. Let $x_i = T_i\Psi_i(x)$, $x_j = T_j\Psi_j(x)$. We have $\sigma(\mathfrak{F}'(x)) = \sigma(\mathfrak{F}'(x_i)) = \sigma(\mathfrak{F}'(x_j))$, that is, $\mathfrak{F}'(x_i) = \mathfrak{F}'(x_j)$, and for $\ell = i,j$

$$\int_{\eta'(\mathfrak{F}'(x_\ell))}^{x_\ell} \tilde{\alpha}_\ell(\mathfrak{F}'(x_\ell)) \equiv \int_{\eta'(\mathfrak{F}'(x))}^{x} \tilde{\alpha}_\ell(\mathfrak{F}'(x))\ \ (\mathrm{mod}(1, \omega_\ell(\sigma(\mathfrak{F}'(x))))).$$

Now

$$\int_{\eta'(\mathfrak{z}'(x))}^{x} \tilde{\alpha}_j(\mathfrak{z}'(x)) = \frac{1}{c_{ij}\omega_i(\sigma\mathfrak{z}'(x))+d_{ij}} \int_{\eta'(\mathfrak{z}'(x))}^{x} \tilde{\alpha}_i(\mathfrak{z}'(x))$$

and

$$\int_{\eta'(\mathfrak{z}'(x_j))}^{x_i} \tilde{\alpha}_j(\mathfrak{z}'(x_j)) = \frac{1}{c_{ij}\omega_i(\sigma\mathfrak{z}'(x))+d_{ij}} \int_{\eta'(\mathfrak{z}'(x_i))}^{x_i} \tilde{\alpha}_i(\mathfrak{z}'(x_i)) =$$

$$= \frac{1}{c_{ij}\omega_i(\sigma\mathfrak{z}'(x))+d_{ij}} \left[\int_{\eta'(\mathfrak{z}'(x))}^{x} \tilde{\alpha}_i(\mathfrak{z}'(x) + n_1'\omega_i(\sigma(\mathfrak{z}'(x))+n_2'\right] =$$

$$= \int_{\eta'(\mathfrak{z}'(x))}^{x} \tilde{\alpha}_j(\mathfrak{z}'(x)) + \int_{\substack{n_1'\delta_1(\mathfrak{z}'(x))+\\ +n_2'\delta_2(\mathfrak{z}'(x))}} \tilde{\alpha}_j(\mathfrak{z}'(x)) \equiv \int_{\eta'(\mathfrak{z}'(x_j))}^{x_j} \tilde{\alpha}_j(\mathfrak{z}'(x_j))(\mathrm{mod}(1,\omega_j(\sigma(\mathfrak{z}'(x))))$$

$(n_1', n_2' \in \mathbb{Z})$.

We see that $x_i = x_j$. Thus $\nu_i(x) = \nu_j(x)$ and we can define a holomorphic map $\nu: Z' \longrightarrow Z_o' \times D_{\epsilon_3'}$ with $\nu|_{\mathfrak{z}'^{-1}(V_i)} = \nu_i$. It is easy to verify that ν is an isomorphism. Taking $i_Z = i'\cdot(\nu^{-1})$ we finish the proof of Lemma 5. Q.E.D.

This also finishes the proof of Lemma 4 (and of Theorem 8).

Q.E.D.

Theorem 8a. Let $f\colon V \longrightarrow \Delta$ be an elliptic fibration with multiple fibers only of type mI_o. Then there exists a commutative diagram of proper surjective holomorphic maps of complex manifolds

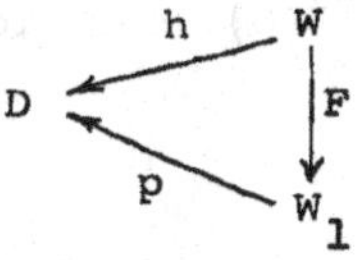

where $D = \{\tau \in \mathbb{C} \,|\, |\tau| < 1\}$ such that $F|_{h^{-1}(o)}\colon h^{-1}(o) \longrightarrow p^{-1}(o)$ coincides with $f\colon V \longrightarrow \Delta$, h has no critical values and for any $\tau \in D-O$, $F|_{h^{-1}(\tau)}\colon h^{-1}(\tau) \longrightarrow p^{-1}(\tau)$ is an elliptic fibration which has singular fibers only of types mI_o and I_1.

Proof. Using Lemma 5 and the same arguments which we used for deduction of Theorem 8 from Lemma 4 (see pp. 118-121) we see that Theorem 8a will be proved if we prove the following

Lemma 6. Let $f'\colon V' \longrightarrow D$ be a proper surjective map of complex manifolds with a single critical value $o \in D$ such that for any $\sigma \in D-o$, $f'^{-1}(\sigma)$ is a non-singular elliptic curve and $f'^{-1}(o)$ is a non-multiple fiber of $f'\colon V' \longrightarrow D$. Then there exist positive numbers ϵ, ϵ_1, $\epsilon < 1$, and a commutative diagram of surjective holomorphic maps of complex manifolds

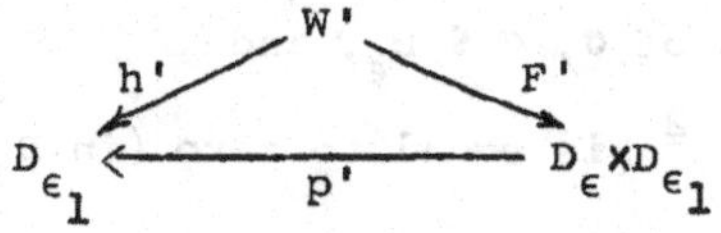

where $D_{\epsilon_1} = \{\tau \in \mathbb{C} \mid |\tau| < \epsilon_1\}$, $D_\epsilon = \{\sigma \in \mathbb{C} \mid |\sigma| < \epsilon\}$, $D_\epsilon \subset D$,

$p': D_\epsilon \times D_{\epsilon_1} \longrightarrow D_{\epsilon_1}$ is the projection, F' is a proper map, such that

a) $F'|_{h'^{-1}(o)}: h'^{-1}(o) \longrightarrow p'^{-1}(o)$ coincides with $f'|_{f'^{-1}(D_\epsilon)}: f'^{-1}(D_\epsilon) \longrightarrow D_\epsilon$;

b) h' has no critical values;

c) For any $\tau \in D_{\epsilon_1} - o$, $F'|_{h'^{-1}(\tau)}: h'^{-1}(\tau) \longrightarrow p'^{-1}(\tau)$ has as generic fiber a non-singular elliptic curve and all singular fibers of $F'|_{h'^{-1}(\tau)} h'^{-1}(\tau) \longrightarrow p'^{-1}(\tau)$ are of type I_1;

d) There exists a holomorphic cross-section η of $F': W' \longrightarrow D_\epsilon \times D_{\epsilon_i}$.

Proof of Lemma 6. Taking ϵ sufficiently small we can assume that $f'|_{f'^{-1}(D_\epsilon)}: f'^{-1}(D_\epsilon) \longrightarrow D_\epsilon$ is obtained by minimal resolution of singularities from the family of elliptic curves which is given in $D_\epsilon \times \mathbb{C}P^2$ by the following equation:

$$z_o z_2^2 = z_1^3 + p(\sigma) z_1 z_o^2 + q(\sigma) z_o^3 ,$$

where $(z_o : z_1 : z_2)$ are homogeneous coordinates in $\mathbb{C}P^2$, $p(\sigma), q(\sigma)$ are holomorphic functions of σ, $\sigma \in D_\epsilon$, and $\Delta(\sigma) = 4(p(\sigma))^3 + 27(q(\sigma))^2$ is equal to zero (in D_ϵ) only at $\sigma = o$ (see [15], Ch. VII, §6).

Suppose for a moment that $(p')^3q + 2(q')^3 \equiv 0$ and $(p')^2p+3(q')^2 \equiv 0$ (where $p' = \frac{dp}{d\sigma}$, $q' = \frac{dq}{d\sigma}$). Then $3(p')^3q - 2(p')^2pq' \equiv 0$ and because $p' \not\equiv 0$ (if not, $q' \equiv 0$ and $p \equiv \text{const}$, $q \equiv \text{const}$, $\Delta \equiv \text{const}$) we have $3p'q-2q'p \equiv 0$ and $q^2 = cp^3$, $c \in \mathbb{C}$. Now $q = \sqrt{c}\, p^{3/2}$, $q' = \sqrt{c}\, \frac{3}{2}p^{1/2}p'$, $(q')^2 = c\,\frac{9}{4}p(p')^2$ and $(p')^2p + 3c\frac{9}{4}p(p')^2 \equiv 0$, $c = -\frac{4}{27}$ (we used $p \not\equiv 0$ (because $p' \not\equiv 0$)). We see that $\Delta(\sigma) \equiv 0$. Contradiction.

We obtain that almost for all $a_o \in \mathbb{C}$, $(p')^3q+2(q')^3-a_op'[(p')^2p+3(q')^2]$ is a non-zero holomorphic function of $\sigma(\in D_\epsilon)$. Choose a_o with such a property and demand also that $a_o \neq 0$, $q(\sigma)-a_op(\sigma) \not\equiv 0$ and $(a_o)^3+p(o)a_o-q(o) \neq 0$.

Now let X be a complex analytic subvariety in $D_\epsilon \times D_{\epsilon'} \times \mathbb{C}P^2$ defined by the following equation:

$$z_oz_2^2 = z_1^3+(p(\sigma)+\tau)z_1z_o^2 + (q(\sigma)+a_o\tau)z_o^3.$$

Suppose that $\tilde{x} = (\tilde{\sigma},\tilde{\tau};\tilde{z}_o:\tilde{z}_1:\tilde{z}_2) \in X$ is either a singular point of X or a critical point of the function τ considered as a holomorphic function on X. Then we have

$$\begin{aligned} &-\tilde{z}_2^2 + 2\tilde{z}_o\tilde{z}_1(p(\tilde{\sigma})+\tilde{\tau}) + 3\tilde{z}_o^2(q(\tilde{\sigma})+a_o\tilde{\tau}) = 0 \\ &3\tilde{z}_1^2 + \tilde{z}_o^2(p(\tilde{\sigma})+\tilde{\tau}) = 0; \\ &-2\tilde{z}_2\tilde{z}_o = 0 \\ &p'(\tilde{\sigma})\tilde{z}_1\tilde{z}_o^2 + q'(\tilde{\sigma})\tilde{z}_o^3 = 0. \end{aligned}$$

We see that $\tilde{z}_o \neq 0$, $\tilde{z}_2 = 0$ and

$$4[p(\tilde{\sigma})+\tilde{\tau}]^3 + 27[q(\tilde{\sigma})+a_o\tilde{\tau}]^2 = 0$$

$$2q'(\tilde{\sigma})[p(\tilde{\sigma})+\tilde{\tau}]-3p'(\tilde{\sigma})[q(\tilde{\sigma})+a_o\tilde{\tau}] = 0.$$

Suppose $p'(\tilde{\sigma}) = 0$, $q'(\tilde{\sigma}) = 0$. Then taking ϵ smaller we can assume that $\tilde{\sigma} = 0$ (because it cannot be $p'(\sigma) \equiv 0$, $q'(\sigma) \equiv 0$). Taking ϵ' smaller, we can assume that the equation $4[p(o)+\tau]^3 + 27[q(o)+a_o\tau]^2 = 0$ has only one zero τ with $|\tau| < \epsilon'$, namely $\tau = 0$. Thus we get $\tilde{\tau} = 0$.

If $p'(\tilde{\sigma}) = 0$, $q'(\tilde{\sigma}) \neq 0$, we obtain $p(\tilde{\sigma})+\tilde{\tau} = 0$. Hence $q(\tilde{\sigma})+a_o\tilde{\tau} = 0$ and $q(\tilde{\sigma})-a_o p(\tilde{\sigma}) = 0$. Taking ϵ smaller, we can assume that the holomorphic function $q(\sigma)-a_o p(\sigma)$ has at most one zero in D_ϵ and at the point $\sigma = 0$ (if this zero exists). Taking ϵ' smaller we can assume that the equatinn $p(o)+\tau = 0$ has at most one zero τ with $|\tau| < \epsilon'$ and this is $\tau = 0$ (if this zero exists, that is, if $p(o) = 0$). Again we get $\tilde{\tau} = 0$.

Consider now the case $p'(\tilde{\sigma}) \neq 0$. Then

$$q(\tilde{\sigma}) + a_o\tilde{\tau} = \frac{2}{3}\frac{q'(\tilde{\sigma})}{p'(\tilde{\sigma})}[p(\tilde{\sigma})+\tilde{\tau}]$$

and

$$4(p(\tilde{\sigma})+\tilde{\tau})^3 + \frac{27\cdot 4}{9}\frac{(q'(\tilde{\sigma}))^2}{(p'(\tilde{\sigma}))^2}(p(\tilde{\sigma})+\tilde{\tau})^2 = 0.$$

If $p(\tilde{\sigma})+\tilde{\tau} = 0$, we obtain as above that $\tilde{\tau} = 0$. Suppose $p(\tilde{\sigma})+\tilde{\tau} \neq 0$. Then

$$p(\tilde{\sigma})+\tilde{\tau} = -3\,\frac{(q'(\tilde{\sigma}))^2}{(p'(\tilde{\sigma}))^2}, \quad q(\tilde{\sigma})+a_o\tilde{\tau} = -2\,\frac{(q'(\tilde{\sigma}))^3}{(p'(\tilde{\sigma}))^3}$$

and

$$q(\tilde{\sigma})-a_o p(\tilde{\sigma}) = -2\,\frac{(q'(\tilde{\sigma}))^3}{(p'(\tilde{\sigma}))^3} + 3a_o\frac{(q'(\tilde{\sigma}))^2}{(p'(\tilde{\sigma}))^2},$$

$$(p'(\tilde{\sigma}))^3 q(\tilde{\sigma})+2[q'(\tilde{\sigma})]^3 - a_o p'(\tilde{\sigma})[(p'(\tilde{\sigma}))^2 p(\tilde{\sigma})+3(q'(\tilde{\sigma}))^2] = 0.$$

Taking ϵ smaller, we can assume that the holomorphic function

$$(p')^3 q + 2(q')^3 - a_o p'[(p')^2 p+3(q')^2]$$

has at most one zero in D_ϵ and at the point $\sigma = 0$ (if this zero exists). As above, we see from the cubic equation

$$4[p(o)+\tau]^3 + 27[q(o)+a_o\tau]^2 = 0$$

that $\tilde{\tau} = 0$.

We proved that X has no singular points with $\tau = 0$ and that the function $\tau|_X$ has no critical values in $D_{\epsilon'}-0$.

Let S be a complex subspace of $D_{\epsilon'}\times D_\epsilon$ defined by

$$\tilde{\Delta}(\sigma,\tau) = 4(p(\sigma)+\tau)^3 + 27(q(\sigma)+a_o\tau)^2 = 0.$$

Suppose that S has a multiple component S_1 with $S_1 \ni (0,0)$. For any $z' = (\sigma',\tau') \in S_1$ we have $\tilde{\Delta}(\sigma',\tau') = 0$, $\frac{\partial}{\partial\tau}\tilde{\Delta}(\sigma',\tau') = 0$.

But

We see that $\tilde{z}_o \neq 0$, $\tilde{z}_2 = 0$ and

$$4[p(\tilde{\sigma})+\tilde{\tau}]^3 + 27[q(\tilde{\sigma})+a_o\tilde{\tau}]^2 = 0$$

$$2q'(\tilde{\sigma})[p(\tilde{\sigma})+\tilde{\tau}]-3p'(\tilde{\sigma})[q(\tilde{\sigma})+a_o\tilde{\tau}] = 0.$$

Suppose $p'(\tilde{\sigma}) = 0$, $q'(\tilde{\sigma}) = 0$. Then taking ϵ smaller we can assume that $\tilde{\sigma} = 0$ (because it cannot be $p'(\sigma) \equiv 0$, $q'(\sigma) \equiv 0$). Taking ϵ' smaller, we can assume that the equation $4[p(o)+\tau]^3 + 27[q(o)+a_o\tau]^2 = 0$ has only one zero τ with $|\tau| < \epsilon'$, namely $\tau = 0$. Thus we get $\tilde{\tau} = 0$.

If $p'(\tilde{\sigma}) = 0$, $q'(\tilde{\sigma}) \neq 0$, we obtain $p(\tilde{\sigma})+\tilde{\tau} = 0$. Hence $q(\tilde{\sigma})+a_o\tilde{\tau} = 0$ and $q(\tilde{\sigma})-a_o p(\tilde{\sigma}) = 0$. Taking ϵ smaller, we can assume that the holomorphic function $q(\sigma)-a_o p(\sigma)$ has at most one zero in D_ϵ and at the point $\sigma = 0$ (if this zero exists). Taking ϵ' smaller we can assume that the equatinn $p(o)+\tau = 0$ has at most one zero τ with $|\tau| < \epsilon'$ and this is $\tau = 0$ (if this zero exists, that is, if $p(o) = 0$). Again we get $\tilde{\tau} = 0$.

Consider now the case $p'(\tilde{\sigma}) \neq 0$. Then

$$q(\tilde{\sigma}) + a_o\tilde{\tau} = \frac{2}{3}\frac{q'(\tilde{\sigma})}{p'(\tilde{\sigma})}[p(\tilde{\sigma})+\tilde{\tau}]$$

and

$$4(p(\tilde{\sigma})+\tilde{\tau})^3 + \frac{27\cdot 4}{9}\frac{(q'(\tilde{\sigma}))^2}{(p'(\tilde{\sigma}))^2}(p(\tilde{\sigma})+\tilde{\tau})^2 = 0.$$

If $p(\tilde{\sigma})+\tilde{\tau} = 0$, we obtain as above that $\tilde{\tau} = 0$. Suppose $p(\tilde{\sigma})+\tilde{\tau} \neq 0$. Then

then for $\tau_o \neq 0$ the family $f_{\tau_o} : X_{\tau_o} \longrightarrow \tau_o \times D_\epsilon$ has only singular fibers of type I_1 (see [15], Ch. VII, §6).

It is well known that X_o has as singularities only a single rational double-point. Taking simultaneous resolution of singularities for the family $\{X_\tau, \tau \in D_{\epsilon'}\}$ (see [6],[7]), we get a diagram

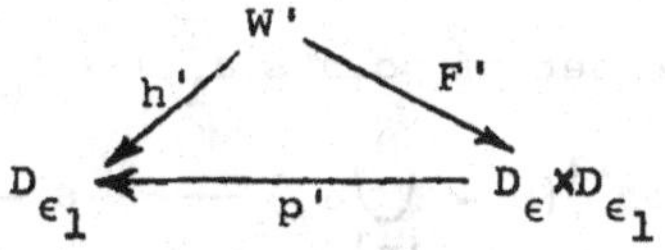

with desired properties a), b), c).

d) is evident (the corresponding holomorphic cross-section comes from the cross-section $\{z_o = 0, z_1 = 0, z_2 = 1\}$ of $f: X \longrightarrow D_{\epsilon'} \times D_\epsilon$). Q.E.D.

§2. Lefshetz fibrations of 2-toruses.

Definition 3. Let $f: M \longrightarrow S$ be a differential map of connected compact oriented differential manifolds (which may have boundaries), $\dim M = 4$, $\dim S = 2$. We say that $f: M \longrightarrow S$ is a Lefshetz fibration if the following is true:

a) $\partial M = f^{-1}(\partial S)$;

b) there is a finite set of points $a_1, \cdots, a_\mu \in S - \partial S$ such that

$$f\Big|_{f^{-1}(S - \bigcup_{i=1}^{\mu} a_i)} : f^{-1}(S - \bigcup_{i=1}^{\mu} a_i) \longrightarrow S - \bigcup_{i=1}^{\mu} a_i$$

is a differential fiber bundle with connected fibers;

c) for any $i \in (1,2,\cdots,\mu)$ $H_2(f^{-1}(a_i),\mathbb{Z}) = \mathbb{Z}$ and there exists a single point $c_i \in f^{-1}(a_i)$ such that

c_1) $(df)_x$ is an epimorphism for any $x \in f^{-1}(a_i) - c_i$,

c_2) there exist neighborhoods B_i of a_i in S, U_i of c_i in M and complex coordinates λ_i in B_i and z_{i1}, z_{i2} in U_i, which define in B_i and U_i the same orientations as global orientations of S and M restricted to B_i and U_i correspondingly, $f(U_i) = B_i$ and $f\big|_{U_i} : U_i \longrightarrow B_i$ is given by the following formula:

$$\lambda_i = z_{i1}^2 + z_{i2}^2 .$$

We shall call $a_1, a_2, \cdots, a_\mu$ critical values of f.

Remark 1. If $a \in S - \bigcup_{i=1}^{\mu} a_i$, the genus of $f^{-1}(a)$ does not depend on a. If this genus is equal to one, we shall call $f: M \longrightarrow S$ "Lefshetz fibration of 2-toruses over S".

Remark 2. If $a_o \in S - \bigcup_{i=1}^{\mu} a_i$, $j \in (1,2,\cdots,\mu)$ and γ is a smooth path in S connecting a_o and a_j and such that $a_i \notin \gamma$, $i = 1,2,\cdots,j-1,j+1,\cdots,\mu$, we define by usual arguments of Lefshetz theory (see [16]), so-called "Lefshetz vanishing cycle" $\delta_j \in H_1(f^{-1}(a_o),\mathbb{Z})$ corresponding to γ_j.

Definition 4. Let $f: M \longrightarrow S$ be a Lefshetz fibration of 2-toruses, T^2 be a 2-torus, $\Omega(T^2)$ be the group of all diffeomorphisms of T^2 preserving orientation, $\alpha: S^1 \longrightarrow \Omega(T^2)$ a differential map (S^1 is a circle), $a_1,\cdots,a_\mu$ be all critical values of $f: M \longrightarrow S$, $a \in \mathrm{int}(S) - \bigcup_{i=1}^{\mu} a_i$, D_a be a small closed 2-disk in $\mathrm{int}(S) - \bigcup_{i=1}^{\mu} a_i$ with the center a. Identify ∂D_a with S^1 and $f^{-1}(D_a)$ with $D_a \times T^2$. Let $N_a = f^{-1}(\partial D_a)$, $\tilde{\alpha}: N_a \longrightarrow \partial(D_a \times T^2)$ be a diffeomorphism canonically defined by α (and by the identification $f^{-1}(D_a) = D_a \times T^2$). Denote $M_{a,\alpha} = \overline{M - f^{-1}(D_a)} \cup_{\tilde{\alpha}} (D_a \times T^2)$ and let $f_{a,\alpha}: M_{a,\alpha} \longrightarrow S$ be a map which on $\overline{M - f^{-1}(D_a)}$ is equal to f and on $D_a \times T^2$ is equal to canonical projection $D_a \times T^2 \longrightarrow D_a$. We call $f_{a,\alpha}: M_{a,\alpha} \longrightarrow S$ a Lefshetz fibration of 2-toruses obtained from $f: M \longrightarrow S$ by α-twisting at the point a.

Definition 5. Let $f_1: M_1 \longrightarrow S_1$, $f_2: M_2 \longrightarrow S_2$ are Lefshetz fibrations. We say that f_1 and f_2 are isomorphic (as Lefshetz fibrations) if there exist orientation-preserving diffeomorphisms $\psi: M_1 \longrightarrow M_2$, $\varphi: S_1 \longrightarrow S_2$ such that $\varphi f_1 = f_2 \psi$.

Remark 3. It is easy to prove that the α-twisting $f_{a,\alpha}: M_{a,\alpha} \longrightarrow S$ can always be defined by some $\alpha_o: S^1 \longrightarrow \Omega_o(T^2)$ where $\Omega_o(T^2)$ is the component of the identity element in $\Omega(T^2)$ and that it depends only on the class of α_o in $\pi_1(\Omega_o(T^2), \mathrm{Id})$. Hence we can assume that in the notations $f_{a,\alpha}, M_{a,\alpha}$ symbol α means an element of $\pi_1(\Omega_o(T^2), \mathrm{Id})$.

Lemma 7. Let $f: M \longrightarrow S$ be a Lefshetz fibration of 2-toruses and $f_{a,\alpha}: M_{a,\alpha} \longrightarrow S$ be some α-twisting of $f: M \longrightarrow S$. Suppose that the canonical homomorphism

$$\pi_1(S - \bigcup_{i=1}^{\mu} a_i, a) \longrightarrow \mathrm{Aut}_+(H_1(f^{-1}(a), \mathbb{Z}))$$

is an epimorphism. Then there exists an isomorphism $\Psi: f \longrightarrow f_{a,\alpha}$ of Lefshetz fibrations $f: M \longrightarrow S$ and $f_{a,\alpha}: M_{a,\alpha} \longrightarrow S$, $\Psi = \{\varphi: S \longrightarrow S, \psi: M \longrightarrow M_{a,\alpha}\}$ such that (i) φ = identity;

(ii) if $\tau_1: f^{-1}(D_a) \xrightarrow[\sim]{} D_a \times T^2$ is a trivialization of $f^{-1}(D_a) \longrightarrow D_a$ which we used in the construction of α-twisting and $\tau_2: f_{a,\alpha}^{-1}(D_a) \xrightarrow[\sim]{} D_a \times T^2$ is the trivialization of $f_{a,\alpha}^{-1}(D_a) \longrightarrow D_a$ which we obtain in the construction of α-twisting, then $\Psi\big|_{f^{-1}(D_a)} = \tau_2^{-1} \cdot \tau_1$.

Proof.*) It is well known that the natural embedding $i: T^2 \longrightarrow \Omega_o(T^2)$ $(i(y)(x) = x+y)$ is a homotopy equivalence (see [17]). Hence we can identify $\pi_1(\Omega_o(T^2), Id)$ with $H_1(T^2, \mathbb{Z})$ and consider α as an element of $H_1(f^{-1}(a), \mathbb{Z})$. Let

$$\gamma: [0,1] \longrightarrow S - \bigcup_{i=1}^{\mu} a_i$$

be a path in $S - \bigcup_{i=1}^{\mu} a_i$ with $\gamma(0) = \gamma(1) = a$ and

$$\gamma_*: H_1(f^{-1}(a), \mathbb{Z}) \longrightarrow H_1(f^{-1}(a), \mathbb{Z})$$

be the canonical automorphism corresponding to γ. There exists an isotopy $\varphi_t: S \longrightarrow S$, $t \in [0,1]$, φ_o = identity, such that for some open

$$U \subset S - D_a,\ U \supset \bigcup_{i=1}^{\mu} a_i,$$

$\varphi_t\big|_U$ = identity (for any $t \in [0,1]$), $\varphi_t(a) = \gamma(t)$ and $\varphi_1(D_a) = D_a$. We see that there exists an isomorphism $\Psi': f_{a,\alpha} \longrightarrow f_{a,\gamma_*(\alpha)}$ of Lefshetz fibrations, $\Psi' = \{\varphi': S \longrightarrow S,\ \psi': M_{a,\alpha} \longrightarrow M_{a,\gamma_*(\alpha)}\}$

*)The proof of Lemma 7 which we give here is based on an idea of D. Mumford.

such that $\varphi' = \varphi_1$

$$(\psi'|_{f^{-1}_{a,\alpha}(U)}) \cdot (j_\alpha|_{f^{-1}(U)}) = j_{\gamma_*(\alpha)}|_{f^{-1}(U)},$$

where

$$j_\alpha : \overline{f^{-1}(S-D_a)} \longrightarrow \overline{f^{-1}_{a,\alpha}(S-D_a)}$$

and

$$j_{\gamma_*(\alpha)} : \overline{f^{-1}(S-D_a)} \longrightarrow \overline{f^{-1}_{a,\gamma_*(\alpha)}(S-D_a)}$$

are canonical identifications. Now we can find an isotopy $\nu_t : M_{a,\gamma_*(\alpha)} \longrightarrow M_{a,\gamma_*(\alpha)}$, ν_o = identity, such that $\forall\, t \in [0,1]$

$$f_{a,\gamma_*(\alpha)} \cdot \nu_t = \varphi_{1-t} \cdot \varphi_1^{-1} \cdot f_{a,\gamma_*(\alpha)} \text{ and } \nu_t|_{f^{-1}_{a,\gamma_*(\alpha)}(U)} = \text{identity.}$$

Denote $\psi''_{\alpha,\gamma} = \nu_1 \cdot \psi'$. We obtain an isomorphism of Lefshetz fibrations $\Psi''_{\alpha,\gamma} : f_{a,\alpha} \longrightarrow f_{a,\gamma_*(\alpha)}$,

$$\Psi''_{\alpha,\gamma} = \{\text{id}: S \longrightarrow S,\ \psi''_{\alpha,\gamma} : M_{a,\alpha} \longrightarrow M_{a,\gamma_*(\alpha)}\},$$

It is clear that

$$(\psi''_{\alpha,\gamma}|_{f^{-1}_{a,\alpha}(U)}) \cdot (j_\alpha|_{f^{-1}(U)}) = j_{\gamma_*(\alpha)}|_{f^{-1}(U)}.$$

Let e_1, e_2 be any free basis of $H_1(f^{-1}(a),\mathbb{Z})$. Since $f_{a,\alpha} = (f_{a,\alpha_1})_{a,\alpha_2}$ for $\alpha = \alpha_1 + \alpha_2$ we have to prove our Lemma only for the case $\alpha = e_1$. Let $\Theta \in \mathrm{Aut}_+(H_1(f^{-1}(a),\mathbb{Z}))$ be an automorphism with $\Theta(e_1) = e_1 + e_2$ and $\tilde{\gamma} : [0,1] \longrightarrow S - \bigcup_{i=1}^{\mu} a_i$ be a closed path in $S - \bigcup_{i=1}^{\mu} a_i$ with $\tilde{\gamma}(0) = a$ and $\tilde{\gamma}_* = \theta$. Take

$$\Psi''_{e_1,\tilde{\gamma}} = \{id: S \longrightarrow S,\ \psi''_{e_1,\tilde{\gamma}}: M_{a,e_1} \longrightarrow M_{a,e_1+e_2}\}$$

constructed above.

Now let $\theta' \in \mathrm{Aut}_+(H_1(f^{-1}(a),\mathbb{Z})$ be an automorphism with $\theta'(e_2) = -e_1$ and $\gamma': [0,1] \longrightarrow S - \bigcup_{i=1}^{\mu} a_i$ be a closed path in $S - \bigcup_{i=1}^{\mu} a_i$ with $\gamma'(0) = a$ and $\gamma'_* = \theta'$. We have the following chain of isomorphisms

$$f_{a,e_1} \simeq f_{a,e_1+e_2} \simeq (f_{a,e_1})_{a,e_2} \simeq (f_{a,e_1})_{a,\gamma'_*(e_2)} =$$

$$(f_{a,e_1})_{a,-e_1} \simeq f_{a,0} \simeq f.$$

Denote by $\Psi: f \longrightarrow f_{a,e_1}$, $\Psi = \{\varphi: S \longrightarrow S,\ \psi: M \longrightarrow M_{a,e_1}\}$ the isomorphism which we obtained. It is clear that φ = identity and that there exists an open subset $U \subset S\text{-}D_a$, $U \supset \bigcup_{i=1}^{\mu} a_i$, such that $\psi|_{f^{-1}(U)} = j_{e_1}|_{f^{-1}(U)}$. Now take $b \in U - \bigcup_{i=1}^{\mu} a_i$, $a' \in \partial D_a$ and let $\delta: [0,1] \longrightarrow S - \bigcup_{i=1}^{\mu} a_i$ be a smooth path in $S - \bigcup_{i=1}^{\mu} a_i$ with $\delta(0) = b$, $\delta(1) = a'$.

Using a trivialization of $f: M \longrightarrow S$ over δ we can improve $\psi: f \to f_{a,e_1}$ such that $\psi|_{f^{-1}(a')} = j_{e_1}|_{f^{-1}(a')}$. Now from

$\psi\big|_{f^{-1}(a')} = j_{e_1}\big|_{f^{-1}(a')}$ it easily follows that we can finally improve ψ such that it will satisfy all demands of our Lemma.

Q.E.D.

Remark to Lemma 7. Let $f: M \longrightarrow S$ be the same as in Lemma 7. Suppose in addition that $\partial S \neq \emptyset$ and that there exists a connected component $\underline{s}$ of ∂S such that the fiber bundle $f\big|_{\underline{s}}: f^{-1}(\underline{s}) \longrightarrow \underline{s}$ is trivial. Let $\underline{\psi}: f^{-1}(\underline{s}) \longrightarrow \underline{s} \times T^2$ be some trivialization of $f\big|_{\underline{s}}$. Let $\underline{\alpha} \in \pi_1(\Omega_o(T^2), \mathrm{Id})$ and $\underline{\tilde{\alpha}}$ be a diffeomorphism $f^{-1}(\underline{s}) \longrightarrow f^{-1}(\underline{s})$ (over $\underline{s}$) corresponding to $\underline{\alpha}$. Then there exists a diffeomorphism $\tilde{\beta}: M \longrightarrow M$ such that $f\tilde{\beta} = \tilde{\beta}$ and $\tilde{\beta}\big|_{f^{-1}(\underline{s})} = \underline{\tilde{\alpha}}$.

Proof. Let $\underline{D}$ be a closed 2-disk with $\partial\underline{D} = \underline{s}$, $\underline{M} = M \cup_{\underline{\psi}} (\underline{D} \times T^2)$, $\underline{S} = S \cup_{\text{identity on } \underline{s}} \underline{D}$, $\underline{f}: \underline{M} \longrightarrow \underline{S}$ be defined by $\underline{f}\big|_M = f$, $\underline{f}\big|_{\underline{D}\times T^2}$ = canonical projection $\underline{D}\times T^2 \longrightarrow \underline{D}$. Let $\underline{a}$ be the center of $\underline{D}$. Using $\underline{\tilde{\alpha}}$ and the given trivialization $\underline{f}^{-1}(\underline{D}) = \underline{D}\times T^2$ we define an $\underline{\alpha}$-twisting $\underline{f}_{\underline{a},\underline{\alpha}}: \underline{M}_{\underline{a},\underline{\alpha}} \longrightarrow \underline{S}$ of $\underline{f}: \underline{M} \longrightarrow \underline{S}$. Let $\Psi: \underline{f} \longrightarrow \underline{f}_{\underline{a},\underline{\alpha}}$,

$$\Psi = \{\mathrm{id}: \underline{S} \longrightarrow \underline{S},\ \psi: \underline{M} \longrightarrow \underline{M}_{\underline{a},\underline{\alpha}}\}$$

be an isomorphism which exists by Lemma 7 and has the property (ii) Denote $\tilde{\beta} = \left(\psi\big|_{\underline{f}^{-1}_{\underline{a},\underline{\alpha}}(S)}\right)^{-1}$. Using $\underline{f}^{-1}(S) = M$ and the identification of $\underline{f}^{-1}_{\underline{a},\underline{\alpha}}(S)$ with M as in Definition 4 we can consider $\tilde{\beta}$ as an autodiffeomorphism of M. Now it is easy to see that $f\tilde{\beta} = \tilde{\beta}$ and $\tilde{\beta}|_{f^{-1}(\underline{s})} = \tilde{\underline{\alpha}}$. Q.E.D.

Lemma 7a. Let $f_1: M_1 \longrightarrow S$, $f_2: M_2 \longrightarrow S$ be two Lefshetz fibrations with the same set of critical values, say $\{a_1, \cdots, a_\mu\}$, which is not empty. Let $a \in S - \bigcup_{i=1}^{\mu} a_i$.

Suppose that $f_1^{-1}(a) = f_2^{-1}(a)$ and the corresponding canonical homomorphisms $\pi_1(S - \bigcup_{i=1}^{\mu} a_i, a) \longrightarrow \text{Aut}_+(H_1(f_1^{-1}(a), \mathbb{Z}))$ and $\pi_1(S - \bigcup_{i=1}^{\mu} a_i, a) \longrightarrow \text{Aut}_+(H_1(f_2^{-1}(a), \mathbb{Z}))$ coincide and are epimorphisms. Then there exists an isomorphism of Lefshetz fibrations

$$\Psi: f_1 \longrightarrow f_2, \quad \Psi = \{\varphi: S \longrightarrow S, \; \psi: M_1 \longrightarrow M_2\},$$

such that $\varphi(a) = a$, φ induces identity on $\pi_1(S - \bigcup_{i=1}^{a_i}, a)$ and $\psi|_{f_1^{-1}(a)} = \text{identity}$.

Proof. It follows from Lemma 7 that it is sufficient to prove the following

Statement. Suppose that ∂S is non-empty and connected, $S = S' \cup D$, where D is a closed 2-disk, $D \cap \partial S$ is a segment in ∂D, D contains exactly one of $a_1, \cdots, a_\mu$, say a_1, S' is a 2-manifold with boundary and $S' \cap D = \partial S' \cap \partial D = \overline{\partial D - \partial D \cap \partial S}$.

Denote by $M_j' = f_j^{-1}(S')$, $f_j' = f_j|_{M_j'}: M_j' \longrightarrow S'$, $j = 1,2$. Suppose that there exists an isomorphism $\Psi': f_1' \longrightarrow f_2'$, $\Psi' = \{\varphi': S' \longrightarrow S', \psi': M_1' \longrightarrow M_2'\}$, such that $\varphi'(a) = a$, φ' induces identity on $\pi_1(S' - \bigcup_{i=2}^{\mu} a_i, a)$ and $\psi'|_{f_1^{-1}(a)} = id$, $\varphi'(S' \cap D) = S' \cap D$. Then there exists an isomorphism

$$\Psi: f_1 \longrightarrow f_2, \quad \Psi = \{\varphi: S \longrightarrow S, \psi: M_1 \longrightarrow M_2\}$$

such that $\varphi|_{S'} = \varphi'$, $\psi|_{S'} = \psi'$ and φ induces identity on $\pi_1(S - \bigcup_{i=1}^{\mu} a_i)$.

<u>Proof of the Statement</u>. Let c_j, $i = 1,2$, be the singular point of $f_j^{-1}(a_1)$. From the definition of a Lefshetz fibration we have that for each $j = 1,2$ there exist neighborhoods B of a_1 in D, U_j of c_j in $f_j^{-1}(D)$ and complex coordinates λ_j in B, z_{j1}, z_{j2} in U_j which define the same orientations as the global orientations of S and M, restricted to B and U_j correspondingly, and are such that $f(U_j) = B$ and $f|_{U_j}: U_j \longrightarrow B$ is given by the following formula: $\lambda_j = z_{j1}^2 + z_{j2}^2$. For a small $\epsilon > 0$ let $D_{\epsilon,j} = \{b \in B \mid \lambda_j(b) \leq \epsilon\}$, $a_j' \in D_{\epsilon,j}$ be defined by $\lambda_j(a_j') = \epsilon$, a_1'' be a point in $\operatorname{int}(\partial S' \cap \partial D)$, $a_2'' = \varphi'(a_1'')$, $\gamma_1: [0,1] \longrightarrow \overline{D - D_{\epsilon,1}}$ be a smooth path with $\gamma_1(0) = a_1''$, $\gamma_1(1) = a_1'$, $\Gamma_1: I \times I \longrightarrow \overline{D - D_{\epsilon,1}}$ be a (smooth) embedding with

$\Gamma_1(o \times I) \subset \mathrm{int}(\partial S' \cap \partial D)$, $\Gamma_1(1 \times I) \subset \partial D_{\epsilon,1}$, $\Gamma_1(I \times \frac{1}{2}) = \gamma_1$.

Then there exists an embedding $\Gamma_2\colon I \times I \longrightarrow \overline{D - D_{\epsilon,2}}$ with $\Gamma_2(o \times I) \subset \mathrm{int}(\partial S' \cap \partial D)$, $\Gamma_2(1 \times I) \subset D_{\epsilon,2}$ and a diffeomorphism

$$\bar{\varphi}\colon S' \cup \Gamma_1(I \times I) \cup D_{\epsilon,1} \longrightarrow S' \cup \Gamma_2(I \times I) \cup D_{\epsilon,2}$$

such that $\bar{\varphi}|_{S'} = \varphi'$, $\bar{\varphi} \cdot \Gamma_1 = \Gamma_2$, $\bar{\varphi}(D_{\epsilon,1}) = D_{\epsilon,2}$ and for any $b \in D_{\epsilon,1}$, $\lambda_2(\bar{\varphi}(b)) = \lambda_1(b)$. Define the Lefshetz vanishing cycle $\delta_j \in H_1(f_j^{-1}(a_j'))$ as the homology class containing the circle

$$\tilde{\delta}_j\colon (\mathrm{Re}\, z_{j1})^2 + (\mathrm{Re}\, z_{j2})^2 = \epsilon, \quad \mathrm{Im}\, z_{j1} = \mathrm{Im}\, z_{j2} = 0,$$

where orientation on $\tilde{\delta}_j$ is taken according to the order $(\mathrm{Re}\, z_{j1}, \mathrm{Re}\, z_{j2})$ of the corresponding real coordinates.

Let $\gamma_2 = \bar{\varphi}(\gamma_1)$, $\delta_j'' = (\gamma_j^{-1})_*(\delta_j)$, $j = 1,2$. From the classical Pickard-Lefshetz formula and from our assumptions about the representations of $\pi_1(S - \bigcup_{i=1}^{\mu} a_i, a)$ in

$$\mathrm{Aut}_+(H_1(f_1^{-1}(a), \mathbb{Z}) = \mathrm{Aut}_+(H_1(f_2^{-1}(a), \mathbb{Z})$$

corresponding to f_1 and f_2 it follows that $(\psi'|_{f_1^{-1}(a_1'')})_*(\delta_1'') = \pm\delta_2''$.

In the case when we have $-\delta_2''$ we change the numeration of z_{21}, z_{22}. Then for the new δ_2'' we will have $(\psi'|_{f_1^{-1}(a_1'')})_*(\delta_1'') = \delta_2''$.

For j = 1,2 and a small $\epsilon' > 0$ denote by

$$R_{j,\epsilon'} = \{x \in U_j \mid |z_{j1}(x)|^2 + |z_{j2}(x)|^2 \leq \epsilon'\}.$$

Taking $\epsilon << \epsilon'$ we can assume that for each j = 1,2 and for any $b \in D_{\epsilon,j}$, $f_j^{-1}(b)$ is transversal to $\partial R_{j,\epsilon'}$. Define a diffeomorphism

$$\tau: f_1^{-1}(D_{\epsilon,1}) \cap R_{1,\epsilon'} \longrightarrow f_2^{-1}(D_{\epsilon,2}) \cap R_{2,\epsilon'}$$

by

$$z_{21}(\tau(x)) = z_{11}(x), \quad z_{22}(\tau(x)) = z_{12}(x).$$

It is clear that $f_2\tau = \bar{\varphi}f_1$ and $\tau(\tilde{\delta}_1) = \tilde{\delta}_2$. Since $f_j^{-1}(a_j')\cap R_{j,\epsilon'}$ is a tubular neighborhood of $\tilde{\delta}_j$ in $f_j^{-1}(a_j')$ and $(\Psi'|_{f_1^{-1}(a_1'')})_*(\delta_1'') = \delta_2''$ we can construct a diffeomorphism $\bar{\Psi}'': f_1^{-1}(S' \cup \Gamma_1(I\times I)) \longrightarrow f_2^{-1}(S' \cup \Gamma_2(I\times I))$ such that $\bar{\Psi}''|_{S'} = \Psi'$,

$$f_2 \circ \bar{\Psi}'' = (\bar{\varphi}|_{S'\cup\Gamma_1(I\times I)}) \cdot (f_1|_{f_1^{-1}(S'\cup\Gamma_1(I\times I))})$$

and

$$\bar{\Psi}''|_{f_1^{-1}(\Gamma_1(1\times I)) \cap R_{1,\epsilon'}} = \tau|_{f_1^{-1}(\Gamma_1(1\times I)) \cap R_{1,\epsilon'}} .$$

Without loss of generality we can assume that

$$\Gamma_1(1\times I) = \{b \in D_{\epsilon,1} \mid |\lambda_1(b)| = \epsilon, -\frac{\pi}{4} \leq \arg \lambda_1(b) \leq \frac{\pi}{4}\}.$$

Then

$$\Gamma_2(1\times I) = \{b \in D_{\epsilon,2} \,\big|\, |\lambda_2(b)| = \epsilon, -\frac{\pi}{4} \le \arg \lambda_2(b) \le \frac{\pi}{4}\}.$$

For $t \in [-\frac{\pi}{4}, \frac{\pi}{4}]$ define $e_{jt} = \{b \in D_{\epsilon,j} \,|\, \operatorname{Im} \lambda_j(b) = \operatorname{Im} e^{it}\}$,

and $Q_j = \bigcup_{t\in[-\frac{\pi}{4},\frac{\pi}{4}]} e_{jt}$, $j = 1,2$. Now choose a Riemannian metric,

g_1, on $f_1^{-1}(D)$ such that for any $b \in D_{\epsilon,1}$, $f_1^{-1}(b)$ will be

orthogonal to $\partial R_{1,\epsilon}$, and construct a family of trajectories

$$q_y\colon [0,1] \longrightarrow \overline{f_1^{-1}(Q_1) - f_1^{-1}(Q_1) \cap R_{1,\epsilon}},$$

where

$$y \in \overline{f_1^{-1}(\Gamma_1(1\times I)) - f_1^{-1}(\Gamma_1(1\times I)) \cap R_{1,\epsilon}},$$

$$f_1 q_y([0,1]) = e_{j\ \arg \lambda_1(f_1(y))},$$

for any $s \in [0,1]$, $q_y([0,1])$ is orthogonal to $f_1^{-1}(f_1(q_y(s)))$ at the point $q_y(s)$ and $f_1(q_y(s))$ depends only on $f_1(y)$. It is clear that for $y \in f_1^{-1}(\Gamma_1(1\times I)) \cap \partial R_{1,\epsilon}$, we have $q_y([0,1]) \subset f_1^{-1}(Q_1) \cap \partial R_{1,\epsilon}$. Now define a family of trajectories $\overline{q}'_{\overline{y}'}\colon [0,1] \longrightarrow f_2^{-1}(Q_2) \cap \partial R_{2,\epsilon}$, $\overline{y}' \in f_2^{-1}(\Gamma_2(1\times I)) \cap \partial R_{2,\epsilon}$, as follows:

$$\overline{q}'_{\overline{y}'} = \tau \cdot q_{\tau^{-1}(\overline{y}')}\ .$$

We can choose a Riemannian metric g_2 on $f_2^{-1}(D)$ such that for any $b \in D_{\epsilon,2}$, $f_2^{-1}(b)$ will be orthogonal to $\partial R_{2,\epsilon}$, and for any

$s \in [0,1]$, $\bar{q}'_{\bar{y}'}([0,1])$ will be orthogonal to $f_2^{-1}(f_2(\bar{q}'_{\bar{y}'}(s)))$ at the point $\bar{q}'_{\bar{y}'}(s)$. Now using g_2 we extend the family $\{\bar{q}'_{\bar{y}'}\}$ to the family of trajectories $q'_{y'}: [0,1] \longrightarrow f_2^{-1}(Q_2)-f_2^{-1}(Q_2)\cap R_{2,\epsilon'}$ with the properties analogous to the properties of the family $\{q_y\}$. Using the families $\{q_y\}$ and $\{q'_{y'}\}$ and the diffeomorphism τ we extend $\bar{\Psi}''$ to a diffeomorphism

$$\bar{\bar{\Psi}}'': f_1^{-1}(S' \cup \Gamma_1(I\times I) \cup Q_1) \longrightarrow f_2^{-1}(S' \cup \Gamma_2(I\times I) \cup Q_2)$$

such that

$$\bar{\bar{\Psi}}''|_{S'\cup\Gamma_1(I\times I)} = \bar{\Psi}'',$$

$$f_2 \cdot \bar{\bar{\Psi}}'' = (\bar{\varphi}|_{S'\cup\Gamma_1(I\times I)\cup Q_1}) \cdot (f_1|_{f_1^{-1}(S'\cup\Gamma_1(I\times I)\cup Q_1)}) .$$

Our statement easily follows from the existence of $\bar{\bar{\Psi}}''$ with these properties. Q.E.D.

Definition 6. We shall say that a Lefshetz fibration of 2-toruses $f: M \longrightarrow S$ is regular if S is diffeomorphic to S^2 (2-dimensional sphere), and the set of critical values of f is not empty.

Definition 7. Let $f_i: M_i \longrightarrow S_i$, $i = 1,2$ be two Lefshetz fibrations, $\partial S_1 = \partial S_2 = \emptyset$, such that the non-singular fibers of f_1 and f_2 have the same genus. Let $a^{(i)} \in S_i$, $i = 1,2$, be some non-critical values of f_i, $D_{a^{(i)}}$ be a closed 2-disk in S_i with the center $a^{(i)}$ which does not contain critical values of f_i, and $\beta: \partial D_{a^{(1)}} \longrightarrow \partial D_{a^{(2)}}$ be some orientation reversing diffeomorphism. Identify $f_i^{-1}(D_{a^{(i)}})$ with $D_{a^{(i)}} \times C$, where C is the typical non-singular fiber of f_1 and f_2. Let $\tilde{\beta}: f_1^{-1}(\partial D_{a^{(1)}}) \longrightarrow f_2^{-1}(\partial D_{a^{(2)}})$ be defined by $\tilde{\beta} = \beta \times (\text{Identity})$ and

$$M = M_1 \oplus M_2 = \overline{(M_1 - f_1^{-1}(D_{a^{(1)}})} \cup_{\tilde{\beta}} \overline{(M_2 - f_2^{-1}(D_{a^{(2)}}))},$$

$$S = S_1 \# S_2 = \overline{(S_1 - D_{a^{(1)}})} \cup_{\beta} \overline{(S_2 - D_{a^{(2)}})},$$

$f: M \longrightarrow S$ be defined by $f\big|_{\overline{M_i - f_i^{-1}(D_{a^{(i)}})}} = f_i\big|_{\overline{M_i - f_i^{-1}(D_{a^{(i)}})}}$. It is clear that $f: M \longrightarrow S$ is a Lefshetz fibration and we call it "direct sum of f_1 and f_2" and write $f = f_1 \oplus f_2$.

Theorem 9. Let $f: M \longrightarrow S$ be a regular Lefshetz fibration of 2-toruses, $f_o: M_o \longrightarrow S_o$ be a Lefshetz fibration obtained from a Lefshetz pencil of cubic curves in $\mathbb{C}P^2$ by blowing up of (nine) base points of this Lefshetz pencil. Let $e(M)$ be the Euler characteristic of M. Then $e(M) > 0$, $e(M) \equiv 0 \pmod{12}$ and $f: M \longrightarrow S$ is isomorphic to the direct sum of $\frac{e(M)}{12}$ copies of $f_o: M_o \to S_o$.

Proof. Let $a_1, \cdots, a_\mu$ be all critical values of $f: M \longrightarrow S$, $a_o \in S - \bigcup_{i=1}^{\mu} a_i$, $C_o = f^{-1}(a_o)$, $\gamma_1, \cdots, \gamma_\mu$ be disjoint smooth paths which connect a_o with $a_1, \cdots, a_\mu$ respectively,

$\gamma_1',\cdots,\gamma_\mu'$ be generators of $\pi_1(S-\bigcup_{i=1}^{\mu} a_i, a_o)$ which correspond to $\gamma_1,\cdots,\gamma_\mu$, $\theta_1,\cdots,\theta_\mu$ be automorphisms of $H_1(C_o,\mathbb{Z})$ corresponding to $\gamma_1',\cdots,\gamma_\mu'$, $\delta_1,\cdots,\delta_\mu \in H_1(C_o,\mathbb{Z})$ be Lefshetz vanishing cycles corresponding to $\gamma_1,\cdots,\gamma_\mu$ and e_1,e_2 be generators of $H_1(C_o,\mathbb{Z})$ with $(e_1.e_2) = 1$. Let D_j, $j = 0,1,\cdots,\mu$, be small closed disjoint 2-disks in S with the center a_j, $j = 0,1,\cdots,\mu$. We can assume that the sets $\gamma_i \cap D_j$, $i \in (1,2,\cdots,\mu)$, $j \in (1,2,\cdots,\mu)$, $i \neq j$, are empty and that each of the sets $\gamma_i \cap \partial D_i$, $\gamma_i \cap \partial D_o$, $i = 1,2,\cdots,\mu$, has only one point. Let $a_{io} = \gamma_i \cap \partial D_o$, $a_i' = \gamma_i \cap \partial D_i$ and $\overline{\gamma}_i$ be the part of γ_i from a_o till a_i'. We can assume that $\gamma_i' = \overline{\gamma}_i \cdot \partial D_i \cdot \overline{\gamma}_i^{-1}$ (∂D_i is oriented as the boundary of D_i) and that the order of points $a_{1o},\cdots,a_{\mu o}$ on ∂D_o where ∂D_o is oriented as the boundary of D_o coincides with the order of their indexes. Now classical Picard-Lefshetz formula tells us that $\theta_i(z) = z-(z.\delta_i)\delta_i$, $z \in H_1(C_o,\mathbb{Z})$, $i = 1,2,\cdots,\mu$. It is clear also that $\theta_1\,\theta_2\,\cdots\,\theta_\mu = \mathrm{Id}$ (we write the multiplication of automorphisms of $H_1(C_o,\mathbb{Z})$ here and further from the left to the right (as the multiplication of corresponding matrixes!) and use the notation $z\theta = \theta(z)$).

Let $\tilde{x}$ (corresp. $\tilde{y}$) be the automorphism of $H_1(C_o,\mathbb{Z})$ defined by $z\tilde{x} = z-(z.e_1)e_1$ (corresp. $z\tilde{y} = z-(z.e_2)e_2$), $z \in H_1(C_o,\mathbb{Z})$. It is well known that each δ_i is a primitive element of $H_1(C_o,\mathbb{Z})$. Hence there exists an automorphism A_i of $H_1(C_o,\mathbb{Z})$ preserving intersection form and such that $e_1A_i = \delta_i$. For any $z \in H_1(C_o,\mathbb{Z})$ we have

$$zA_i^{-1}\tilde{x}A_i = (zA_i^{-1}-(zA_i^{-1}.e_1)e_1)A_i = z-(zA_i^{-1}.e_1)e_1A_i =$$
$$= z-(zA_i^{-1}A_i.e_1A_i)e_1A_i = z - (z.\delta_i)\delta_i.$$

Thus $\Theta_i = A_i^{-1}\tilde{x}A_i$ and $\prod_{i=1}^{\mu}(A_i^{-1}\tilde{x}A_i) = \mathrm{Id}.$

For $i = 1,2,\cdots,\mu-1$ let $q_i\colon \pi_1(S - \bigcup_{i=1}^{\mu} a_i, a_o) \longrightarrow \pi_1(S - \bigcup_{i=1}^{\mu} a_i, a_o)$ be an automorphism of $\pi_1(S - \bigcup_{i=1}^{\mu} a_i, a_o)$ defined by

$$q_i(\gamma_1') = \gamma_1',\cdots,q_i(\gamma_{i-1}') = \gamma_{i-1}',\ q_i(\gamma_i') = \gamma_{i+1}',$$
$$q_i(\gamma_{i+1}') = \gamma_{i+1}'^{-1}\gamma_i'\gamma_{i+1}',\ q_i(\gamma_{i+2}') = \gamma_{i+2}',\cdots,q_i(\gamma_\mu') = \gamma_\mu'.$$

It is easy to see that $q_i(\gamma_1'),\cdots,q_i(\gamma_\mu')$ correspond to some new choice of disjoint smooth paths $\gamma_1,\cdots,\gamma_\mu$ connecting a_o with $a_1,\cdots,a_\mu$ and the same is true for $q_i^{-1}(\gamma_1'),\cdots,q_i^{-1}(\gamma_\mu')$ (see Figs. 8 and 9). We denote these new paths by $q_i(\gamma_1),\cdots,q_i(\gamma_\mu)$ (corresp. $q_i^{-1}(\gamma_1),\cdots,q_i^{-1}(\gamma_\mu)$. We call transformations q_i and q_i^{-1} elementary transformations of the paths on the base.

Now if G is a group and $(x_1,\cdots,x_\mu)$ is a μ-tuple of elements of G, we call the transformations

$$(x_1,\cdots,x_\mu) \longrightarrow (x_1,\cdots,x_{i-1},x_{i+1},x_{i+1}^{-i}x_ix_{i+1},x_{i+2},\cdots,x_\mu)$$

Fig. 8 (For q_i)

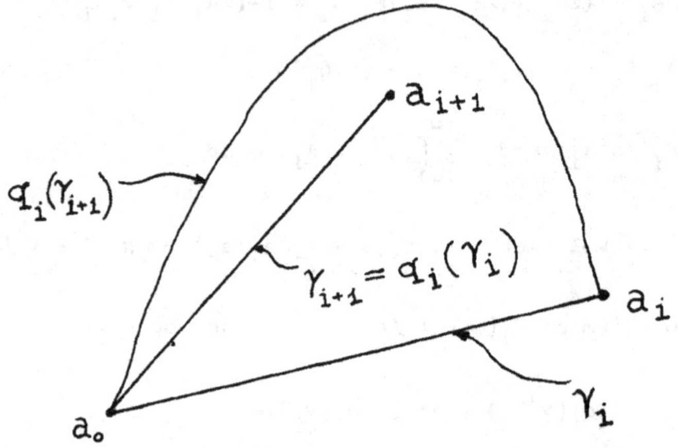

Fig. 9 (For q_i^{-1})

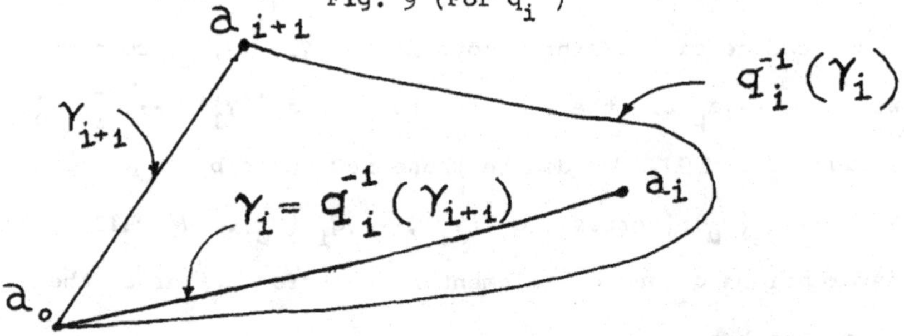

and

$$(x_1,\cdots,x_\mu) \longrightarrow (x_1,\cdots,x_{i-1},x_i x_{i+1} x_i^{-1},x_i,x_{i+2},\cdots,x_\mu)$$

elementary transformations of μ-tuples in G.

Using e_1,e_2 as a basis we identify $H_1(C_o,\mathbb{Z})$ with $\mathbb{Z}\oplus\mathbb{Z}$ and the group of all automorphisms of $H_1(C_o,\mathbb{Z})$ preserving intersection form with the group $SL(2,\mathbb{Z})$. It is clear that an elementary transformation of the paths on the base correspond to some elementary transformation of the μ-tuple $(\theta_1,\cdots,\theta_\mu) \subset SL(2,\mathbb{Z})$ and vice-versa.

We see that if $(\underline{\theta}_1,\cdots,\underline{\theta}_\mu) \subset SL(2,\mathbb{Z})$ is obtained from $(\theta_1,\cdots,\theta_\mu)$ by some finite sequence of elementary transformations then there exists a set of disjoint smooth paths $\underline{\gamma}_1,\cdots,\underline{\gamma}_\mu$ on S connecting a_o with $a_1,\cdots,a_\mu$ and such that $\underline{\theta}_1,\cdots,\underline{\theta}_\mu$ correspond to $\underline{\gamma}_1,\cdots,\underline{\gamma}_\mu$ by the same way as $\theta_1,\cdots,\theta_\mu$ correspond to $\gamma_1,\cdots,\gamma_\mu$.

Now we shall use the following

<u>Lemma 8</u>. Let $\tilde{x} = \left\|\begin{matrix}1 & 0\\ 1 & 1\end{matrix}\right\|$, $\tilde{y} = \left\|\begin{matrix}1 & -1\\ 0 & 1\end{matrix}\right\|$ and $A_1,\cdots,A_\mu \in SL(2,\mathbb{Z})$ be such that if $\theta_i = A_i^{-1}\tilde{x}A_i$, $i = 1,2,\cdots,\mu$, then $\theta_1\cdot\ldots\cdot\theta_\mu = \left\|\begin{matrix}1 & 0\\ 0 & 1\end{matrix}\right\|$. Then $\mu \equiv 0 \pmod 2$ and there exists a finite sequence of elementary transformations starting with some elementary transformation of $(\theta_1,\cdots,\theta_\mu)$ such that if $(\underline{\theta}_1,\cdots,\underline{\theta}_\mu)$ is the resulting μ-tuple in $SL(2,\mathbb{Z})$ then

$$\underline{\theta}_1 = \tilde{x},\ \underline{\theta}_2 = \tilde{y},\cdots,\theta_{2\ell-1} = \tilde{x},\ \theta_{2\ell} = \tilde{y},\cdots,\theta_{\mu-1} = \tilde{x},\ \theta_\mu = \tilde{y}.$$

Proof. We use the following Theorem of R. Livne. (For the proof see Appendix II, page 223.)

Theorem of R. Livne: Let G be a group with two generators a and b and relations $a^3 = b^2 = 1$ (that is, G is isomorphic to the free-product $[(\mathbb{Z}/2\mathbb{Z})*(\mathbb{Z}/3\mathbb{Z})]$). Let $s_i = a^{2-i}ba^i$ for $i = 0,1,2$, and $Q_1,Q_2,\cdots,Q_\mu \in G$ be such that if $y_i = Q_i^{-1}s_oQ_i$, $i = 1,2,\cdots,\mu$, then $y_1 \cdot y_2 \cdot \ldots \cdot y_\mu = 1$. Then there exists a finite sequence of elementary transformations starting with some elementary transformation of $(y_1,\cdots,y_\mu)$ such that if $(\underline{y}_1,\cdots,\underline{y}_\mu)$ is the resulting subset of G then each $\underline{y}_i$, $i = 1,2,\cdots,\mu$, is equal to one of the elements s_o,s_1,s_2.

Complement to the Theorem of R. Livne

Let G,a,b,s_o,s_1,s_2 be the same as above and $y_1,\cdots,y_\mu \in G$ be such that each of y_i, $i = 1,2,\cdots,\mu$, is equal to one of the elements s_o,s_1,s_2 and $y_1 \cdot \ldots \cdot y_\mu = 1$. Then $\mu \equiv 0 \pmod 2$ and there exists a finite sequence of elementary transformations starting with some elementary transformation of $(y_1,y_2,\cdots,y_\mu)$ such that if $(\underline{y}_1,\cdots,\underline{y}_\mu)$ is the resulting μ-tuple in G then for any $j = 1,2,\cdots,\frac{\mu}{2}$, $\underline{y}_{2j-1} = s_1$, $\underline{y}_{2j} = s_2$.

Proof. Note that a cyclic shift in a tuple $(x_1,\cdots,x_\mu)$ can be obtained by a finite sequence of elementary transformations (see Appendix II).

Denote by λ the number of y_j in $y_1,\cdots,y_\mu$ with $y_j = s_o$. Let us call the statement of the Complement to the Theorem of R. Livne in the case when μ,λ are given "Statement $[\mu,\lambda]$". Now consider the set $\mathfrak{B}$ of all pairs $[\tilde{\mu},\tilde{\lambda}]$, $\tilde{\mu} \in \mathbb{Z}$, $\tilde{\lambda} \in \mathbb{Z}$, $\tilde{\mu} > 0$, $\tilde{\lambda} \geq 0$, and define the following order in $\mathfrak{B}$: We say $[\tilde{\mu}_1,\tilde{\lambda}_1] < [\tilde{\mu},\tilde{\lambda}]$ if either $\tilde{\mu}_1 < \tilde{\mu}$ or $\tilde{\mu}_1 = \tilde{\mu}$, $\tilde{\lambda}_1 < \tilde{\lambda}$. Let $\mathfrak{B}'$ be the subset of $\mathfrak{B}$ consisting of all $[\tilde{\mu},\tilde{\lambda}]$ for which the "Statement $[\tilde{\mu},\tilde{\lambda}]$ is not true. Suppose that $\mathfrak{B}' \neq \emptyset$ and let $[\mu,\lambda]$ be a minimal element of $\mathfrak{B}'$.

Suppose that $\lambda \neq 0$. Then we can write $y_1\cdot\ldots\cdot y_\mu$ in the following form

$$y_1\cdot\ldots\cdot y_\mu = s_o^{m_1} X_1 s_o^{m_2} X_2 \ldots . s_o^{m_{\mu'}} X_{\mu'},$$

where

$$(1) \qquad X_j = \prod_{i=0}^{N_j} s_1^{k_{i,j}} s^{\ell_{i,j}}, \quad k_{o,j} \geq 0, \ \ell_{N_j,j} \geq 0 \text{ and if } N_j \neq 0$$

then for all $i = 1,2,\cdots,N_j$, $k_{i,j} > 0$, $\ell_{i-1,j} > 0$

$$(\mu = \sum_{j=1}^{\mu'} m_j + \sum_{j=1}^{\mu'} \sum_{i=0}^{N_j} (k_{i,j}+\ell_{i,j}) \ , \quad \lambda = \sum_{j=1}^{\mu'} m_j).$$

We can write also

$$y_1 \ldots y_\mu = X_1 s_o^{m_2} X_2 \ldots X_{\mu'} s_o^{m_1} .$$

Suppose that $k_{o,j} = 0$. Then because $s_o s_2 = s_2(s_2^{-1} s_o s_2)$ and $s_2^{-1} s_o s_2 = s_1$ we can reduce (by an elementary transformation) our situation to the "Statement $[\mu,\lambda-1]$" which is true. This contradicts $[\mu,\lambda] \in \mathfrak{B}'$. The same arguments show that $\ell_{N_{j,j}} > 0$ (we use $s_1 s_o = (s_1 s_o s_1^{-1}) s_1$ and $s_1 s_o s_1^{-1} = s_2$).

Using elementary transformations on the (ordered) set of factors of X_j, $j = 1,2,\cdots,\mu'$, we can get different expressions for X_j, $j = 1,2,\cdots,\mu'$, as a positive word written in the letters s_1, s_2. Among all these expressions we can find a maximal (for each X_j, $j = 1,2,\cdots,\mu'$) according to lexicographical order (s_1 is the first letter and s_2 is the second one). This shows that we can assume that we already have $y_1,\cdots,y_\mu$ such that each X_j, $j = 1,2,\cdots,\mu$, cannot be transformed by elementary transformations of its factors into a word which in the form (1) (that is, when we put corresponding s_1 and s_2 instead of y_i) is greater according to lexicographical order than X_j.

Take $j \in (1,2,\cdots,\mu')$ with $N_j \neq 0$. Now suppose that some $k_{ij} = 1$, $i \in (1,2,\cdots,N_j)$. We have that in the expression (1) of X_j there is a subproduct which has the following form: $s_2 s_1 s_2$. Using $s_1 s_2 s_1 = s_2 s_1 s_2$, that is, $s_1 = s_2^{-1} s_1^{-1} s_2 s_1 s_2$, we get

$s_2s_1s_2 = s_1(s_1^{-1}s_2s_1)s_2 = s_1s_2[s_2^{-1}(s_1^{-1}s_2s_1)s_2] = s_1s_2s_1.$

It follows from this that using elementary transformations we can replace $s_2s_1s_2$ by $s_1s_2s_1$. This contradicts the maximality of X_j in the lexicographical ordering. Thus we obtain

$k_{i,j} > 1, \quad i = 1,2,\cdots,N_j.$

Suppose that $\ell_{o,j} = 1$. We have $X_j = s_1^{k_{o,j}}s_2s_1\ldots\ldots$. Using substitutions $s_1s_2s_1 \longrightarrow s_2s_1s_2$ we can transform $s_1^{k_{o,j}}s_2s_1$ in $s_2s_1s_2^{k_{o,j}}$. As we showed above, any substitution of the form $s_1s_2s_1 \longrightarrow s_2s_1s_2$ is a sequence of elementary transformations. Thus we come to the situation where $X_j = s_2\ldots\ldots$, but this leads to the contradiction as we explained above. Thus we obtain $\ell_{o,j} > 1$.

Now suppose that for some $i \in (2,3,\ldots,N_j-1)$, $\ell_{i-1,j} = \ell_{i,j} = 1$. We have that in the expression (1) of X_j there is a subproduct which has the following form: $s_1s_2s_1^{k_{i,j}}s_2s_1$. Using substitutions $s_1s_2s_1 \longrightarrow s_2s_1s_2$ and $k_{ij} \geq 2$ we transform $s_1s_2s_1^{k_{i,j}}s_2s_1$ in $s_1s_2s_1s_2s_1s_2s_2^{k_{i,j}-2}$. Hence performing elementary transformations, we can come to the situation when a part of X_j has the form $s_1s_2s_1s_2s_1s_2$. But $s_1s_2s_1s_2s_1s_2 = 1$. This shows that using a cyclic shift we can reduce our situation to the Statement $[\mu-6,\lambda]$ which is true. This contradicts $[\mu,\lambda] \in \mathfrak{B}'$. We see that if for some $i \in (2,3,\cdots,N_j-1)$ $\ell_{i-1,j} = 1$ then $\ell_{i,j} > 1$. Let us prove now that if for some $i \in (2,3,\cdots,N_j)$, $\ell_{i-1,j} = 1$ then $k_{i-1,j} \geq 3$ and $k_{i,j} \geq 3$.

Suppose $k_{i-1,j} = 2$. Then in the expression (1) for X_j we have a part of the following form: $s_2 s_1^2 s_2 s_1^2$. Using elementary transformations we transform this part to $s_1 s_2 s_1 s_2 s_1 s_2$. This contradicts $[\mu,\lambda] \in \mathfrak{B}'$ (as we saw above).

Now suppose that $k_{i,j} = 2$. In this case we have in the expression (1) for X_j a part of the form $s_1^2 s_2 s_1^2 s_2$ which also can be transformed by elementary transformations in $s_1 s_2 s_1 s_2 s_1 s_2$. As above, we get a contradiction with $[\mu,\lambda] \in \mathfrak{B}'$. For $i = 1,2,\cdots,N_j$, we denote $Y_{i,j} = s_2^{\ell_{i-1,j}} s_1^{k_{i,j}}$. We shall say that $Y_{i,j}$ is of the first (corresp. second) kind if $\ell_{i-1,j} > 1$ (corresp. $\ell_{i-1,j} = 1$). We proved that $Y_{1,j}$ is of the first kind and that if $Y_{i,j}$ is of the second kind then $Y_{i-1,j}$ is of the first kind. It follows from this that we can find a set $Z_{1,j},\cdots,Z_{t_j,j}$ of elements of G such that (i) each $Z_{\ell,j}$, $\ell = 1,\cdots,t_j$, is equal either to some $Y_{i,j}$, $i \in (1,2,\cdots,N_j)$ where $Y_{i,j}$ is of the first kind or to some product $Y_{i-1,j} \cdot Y_{i,j}$, $i \in (2,3,\cdots,N_j)$ where $Y_{i,j}$ is of the second kind, and

$$s_1^{k_{o,j}} \prod_{\ell=1}^{t_j} Z_{\ell,j} s_2^{\ell_{N,j}} = X_j.$$

For any element $c \in G$ we define a reduced form of c by writing c as a positive word in the letters a,b (generators of G) and performing all possible cancellations (using $a^3 = b^2 = 1$). It is clear that c has unique reduced form.

Let $Z_{\ell,j}$ be such that $Z_{\ell,j}$ is equal to some $Y_{i,j}$ where $Y_{i,j}$, $i \in (1,\cdots,N_j)$, is of the first kind. Then it is easy to verify that the reduced form of $Z_{\ell,j}$ can be written as $bR_{\ell,j}ba$ with some $R_{\ell,j} \in G$ (we use $s_2^{\ell_{i-1,j}} s_1^{k_{i,j}} = s_2^{\ell_{i-1,j}-2} bs_1^{k_{i,j}-1}$, $\ell_{i-1,j} \geq 2$, $k_{i,j} \geq 2$ and $s_2 = ba^2$, $s_1 = aba$). Consider $Z_{\ell,j}$ which is equal to some $Y_{i-1,j}Y_{i,j}$, $i \in (2,3,\cdots,N_j)$ where $Y_{i,j}$ is of the second kind and $Y_{i-1,j}$ is of the first kind. We have $\ell_{i-2,j} \geq 2$, $\ell_{i-1,j} = 1$ and as we proved above, $k_{i-1,j} \geq 3$, $k_{i,j} \geq 3$. Now we can write

$$Z_{\ell,j} = Y_{i-1,j}Y_{i,j} = s_2^{\ell_{i-2,j}} s_1^{k_{i-1,j}} s_2 s_1^{k_{i,j}}$$

$$= s_2^{\ell_{i-2,j}-2} b\, s_1^{k_{i-1,j}-3} ab\, s_1^{k_{i,j}-3} aba\ .$$

It is easy to see from that formula that the reduced form of $Z_{\ell,j}$ can be written as $(bR_{\ell,j}ba)$ with some $R_{\ell,j} \in G$.

We define $$\pi_j = \begin{cases} 1 & \text{if } N_j = 0 \\ \prod_{\ell=1}^{t_j} (bR_{\ell,j}ba). & \end{cases}$$

It is clear that if $N_j \neq 0$, then the reduced form of $\prod_j$ is equal to $\prod_{\ell=1}^{t_j}(b\, R_{\ell,j} ba)$ and could be written as $bR_j ba$ with some $R_j \in G$. We have $X_j = s_1^{k_{o,j}} \prod_j s_2^{\ell_{N_j,j}}$. If $N_j \neq 0$, that is, $\prod_j \neq 1$, we cannot have cancellations in $s_1^{k_{o,j}} \prod_j s_2^{\ell_{N_j,j}}$ between s_1 and $\prod_j$ and between $\prod_j$ and s_2. If $N_j = 0$, that is, $\prod_j = 1$, we cannot have cancellations in $s_1^{k_{o,j}} s_2^{\ell_{o,j}}$ between s_1 and s_2.

Now $s_o X_j = a^2 babas_1^{k_{o,j}-1} \prod_j s_2^{\ell_{N_j,j}}$,

$$X_j s_o = s_1^{k_{o,j}} \prod_j s_2^{\ell_{N_j,j}-1} ba^2 \cdot a^2 b =$$

$$= s_1^{k_{o,j}} \prod_j s_2^{\ell_{N_j,j}-1} bab.$$

These formulas show that we have no possibilities for further cancellations in $s_o X_j$ and $X_j s_o$.

This contradicts the equality

$$\prod_{i=1}^{\mu} y_i = 1.$$

Now let us consider the case $\lambda = 0$, that is, all $y_1, \cdots, y_\mu$ are equal either to s_1 or to s_2. We can write $X = \prod_{\ell=1}^{\mu} y_i (= 1)$ in the following form:

$$X = \prod_{i=0}^{N} s_1^{k_i} s_2^{\ell_i} \tag{2}$$

where

$$\sum_{i=0}^{N} (k_i + \ell_i) = \mu.$$

Using a cyclic shift of indexes $(1,2,\cdots,\mu)$ we can assume that $k_i > 0$, $\ell_i > 0$ for all $i = 0,1,\cdots,N$. It is clear that $N > 0$. By the same arguments as above we can prove here that X can be written in the form $s_1^{k_0}\Pi s_2^{\ell_N}$ where the reduced form of Π is equal to bRba. HEnce $X = s_1^{k_0}bRbas_2^{\ell_N}$ and we have no possibilities for further cancellations in X. This contradicts $X = 1$. Q.E.D.

Let us return now to the proof of Lemma 8. Let Z be the center of $SL(2,\mathbb{Z})$, $PSL(2,\mathbb{Z}) = SL(2,\mathbb{Z})/Z$, $\varphi\colon SL(2,\mathbb{Z}) \longrightarrow PSL(2,\mathbb{Z})$ be the canonical homomorphism, $x = \varphi(\tilde{x})$, $y = \varphi(\tilde{y})$, $\bar{\theta}_i = \varphi(\theta_i)$, $i = 1,2,\cdots,\mu$, $a = yx$, $b = y^2x$. Clearly $x = aba$, $y = ba^2$. It is well known that $PSL(2,\mathbb{Z})$ is generated by a and b and that all corresponding relations are generated by the relations $a^3 = 1$ and $b^2 = 1$. Since $\bar{\theta}_1\cdot\ldots\cdot\bar{\theta}_\mu = 1$ and $\bar{\theta}_i = \varphi(A_i)^{-1}x\varphi(A_i)$ we can apply the Theorem of R. Livne, and the Complement to it. Thus we get that $\mu \equiv 0 \pmod 2$ and there exists a finite sequence of elementary transformations starting with some elementary transformation of $(\bar{\theta}_1,\cdots,\bar{\theta}_\mu)$ such that if $(\underline{\bar{\theta}}_1,\cdots,\underline{\bar{\theta}}_\mu)$ is the resulting subset of $PSL(2,\mathbb{Z})$ then

$$\underline{\bar{\theta}}_1 = x,\ \underline{\bar{\theta}}_2 = y,\ \cdots,\ \underline{\bar{\theta}}_{2\ell-1} = x,\ \underline{\bar{\theta}}_{2\ell} = y,\ \cdots,\ \underline{\bar{\theta}}_{\mu-1} = x,$$
$$\underline{\bar{\theta}}_\mu = y.$$

It is clear that we can lift our elementary transformations to the group $SL(2,\mathbb{Z})$. Thus we get that there exists a finite sequence of elementary transformations starting with some elementary transformation of $(\theta_1,\cdots,\theta_\mu)$ such that if $(\underline{\theta}_1,\cdots,\underline{\theta}_\mu)$ is the resulting subset of $SL(2,\mathbb{Z})$ then

$$\varphi(\underline{\theta}_1) = x,\ \varphi(\underline{\theta}_2) = y,\cdots,\varphi(\underline{\theta}_{2\ell-1}) = x,\ \varphi(\underline{\theta}_{2\ell}\ \ ) = y,\cdots,$$
$$\varphi(\underline{\theta}_{\mu-1}) = x,\ \varphi(\underline{\theta}_\mu) = y.$$

From the definition of elementary transformation it follows that each $\underline{\theta}_i$, $i = 1,2,\cdots,\mu$, is conjugate to $\tilde{x}$ (and to $\tilde{y}$, because $\tilde{x}$ is conjugate to $\tilde{y}$). A direct verification shows that if θ is an element of $SL(2,\mathbb{Z})$ which is conjugate to $\tilde{x}$ (corresp. to $\tilde{y}$) and $\varphi(\theta) = x$ (corresp. $\varphi(\theta) = y$) then $\theta = \tilde{x}$ (corresp. $\theta = \tilde{y}$). We obtain that for any $j = 1,2,\cdots,\frac{\mu}{2}$, $\underline{\theta}_{2j-1} = \tilde{x}$, $\theta_{2j} = \tilde{y}$.

Q.E.D.

Now we return to the proof of Theorem 9. From Lemma 8 we get that we can assume that paths $\gamma_1,\cdots,\gamma_\mu$ are chosen so that for any $j = 1,2,\cdots,\frac{\mu}{2}$, $\delta_{2j-1} = e_1$, $\delta_{2j} = e_2$. A direct verification shows that if μ_1 is a positive integer with $\mu_1 \leq \mu$, $\mu_1 < 12$, then $\theta_1\cdot\ldots\cdot\theta_{\mu_1} \neq 1$. It is easy to verify that $\theta_1\cdot\ldots\cdot\theta_{12} = 1$. We see that $\mu \equiv 0 \pmod{12}$. Let $k = \frac{\mu}{12}$. Consider $f_o\colon M_o \longrightarrow S_o$ and let $\overline{f}\colon \overline{M} \longrightarrow \overline{S}$ be a Lefshetz fibration which is the direct sum of k copies of $f_o\colon M_o \longrightarrow S_o$.

Let $\{\bar{a}_1,\cdots,\bar{a}_\mu\}$ be the set of critical values of $\bar{f}$, $\bar{a}_o \in \bar{S} - \bigcup_{i=1}^{\mu}\bar{a}_i$. Applying to $\bar{f}: \bar{M} \longrightarrow \bar{S}$ our above considerations we construct a system of disjoint smooth paths $\bar{\gamma}_1,\cdots,\bar{\gamma}_\mu$ connecting $\bar{a}_o$ with $\bar{a}_1,\cdots,\bar{a}_\mu$ such that corresponding Pickard-Lefshetz transformations are given by the matrices $\bar{\Theta}_1,\cdots,\bar{\Theta}_\mu$ with $\bar{\Theta}_{2j-1} = \tilde{x}$, $\bar{\Theta}_{2j} = \tilde{y}$, $j = 1,2,\cdots,\frac{\mu}{2}$, in some free basis $\bar{e}_1,\bar{e}_2$ of $H_1(\bar{f}^{-1}(\bar{a}_o),\mathbb{Z})$.

Now we get from Lemma 7a that $\bar{f}$ is isomorphic to f.

Q.E.D.

Corollary 1. Let $f: M \longrightarrow S$ be a Lefshetz fibration of 2-toruses, $\partial S = \emptyset$. Then M is simply-connected if and only if $f: M \longrightarrow S$ is regular.

Proof. a) Suppose $f: M \longrightarrow S$ is regular. Let $(a_1,\cdots,a_\mu)$ be the set of critical values of $f: M \longrightarrow S$,

$$S' = S-\bigcup_{i=1}^{\mu}a_i,\ M' = f^{-1}(S'),\ f' = f|_{M'}: M' \longrightarrow S',\ a_o \in S',$$

$$C_o = f^{-1}(a_o).$$

From Theorem 9 it follows that there exists a set of smooth disjoint paths $\gamma_1,\cdots,\gamma_\mu$ in S connecting a_o with $a_1,\cdots,a_\mu$ such that the corresponding Lefshetz vanishing cycles $\delta_1,\cdots,\delta_\mu \in H_1(C_o,\mathbb{Z})$ generate $H_1(C_o,\mathbb{Z})$. That means that the image of $\pi_1(C_o)$ in $\pi_1(M)$ is trivial. Since $f'\colon M' \longrightarrow S'$ is a fibre bundle we have the following exact sequence:

(3) $$\pi_1(C_o) \longrightarrow \pi_1(M') \xrightarrow[\beta]{} \pi_1(S').$$

Let D_i, $i = 1,2,\cdots,\mu$, be a small closed 2-disk in S with the center a_i, $s_i = \partial D_i$. From the definition of Lefshetz fibration we get that there exists a local cross-section $\tilde{D}_i$ of $f|_{f^{-1}(D_i)}\colon f^{-1}(D_i) \longrightarrow D_i$. Let $\tilde{s}_i = \partial\tilde{D}_i$. Connecting each $\tilde{s}_i$ with a point $x_o \in C_o$ by some smooth path in M' we get a collection $\{\sigma_1,\cdots,\sigma_\mu\}$ of elements of $\pi_1(M',x_o)$ such that $\beta(\sigma_1),\cdots,\beta(\sigma_\mu)$ generate $\pi_1(S,a_o)$ and images of $\sigma_1,\cdots,\sigma_\mu$ in $\pi_1(M,x_o)$ are trivial. Clearly $\pi_1(M') \longrightarrow \pi_1(M)$ is an epimorphism. Hence we see from the exact sequence (3) that $\pi_1(C_o) \longrightarrow \pi_1(M)$ is an epimorphism. Thus $\pi_1(M) = 0$.

b) Suppose that $\pi_1(M) = 0$. If $\pi_1(S) \neq 0$ we take any $\sigma \in \pi_1(S)$, $\sigma \neq 0$. Then there exists a $\tilde{\sigma} \in \pi_1(M)$ such that $f_*(\tilde{\sigma}) = \sigma$ and $\pi_1(M) \neq 0$. Contradiction. Thus $\pi_1(S) = 0$ and S is diffeomorphic to 2-sphere. Suppose that the set of critical values of $f\colon M \longrightarrow S$ is empty. Then $f\colon M \longrightarrow S$ is a fiber bundle and denoting by C_o its typical fiber we have an exact sequence:

$$\pi_2(S) \longrightarrow \pi_1(C_o) \longrightarrow \pi_1(M) \longrightarrow \pi_1(S).$$

But $\pi_1(S) = 0$, $\pi_2(S) = \mathbb{Z}$, $\pi_1(C_0) = \mathbb{Z} \oplus \mathbb{Z}$. Hence $\pi_1(M) \neq 0$. Contradiction. Q.E.D.

Corollary 2. Let $f_1: M_1 \longrightarrow S_1$, $f_2: M_2 \longrightarrow S_2$ be two regular Lefshetz fibrations. Then f_1 and f_2 are isomorphic (as Lefshetz fibrations) if and only if the corresponding two-dimensional Betti numbers $b_2(M_1)$ and $b_2(M_2)$ are equal.

Proof. Immediately follows from Theorem 9. Q.E.D.

The next Corollary gives another approach to a result of A. Kas (see [18]).

Corollary 3 (A. Kas). Let V_1 and V_2 be elliptic surfaces over $\mathbb{C}P^2$ with no multiple fibers, with at least one singular fibre and with no exceptional curve contained in a fiber. Then V_1 and V_2 are diffeomorphic if and only if $b_2(V_1) = b_2(V_2)$.

Proof. It follows from Theorem 8a that we can assume that V_1 and V_2 have only singular fibers of type I_1. Then the corresponding maps $f_1: V_1 \longrightarrow \mathbb{C}P^1$, $f_2: V_2 \longrightarrow \mathbb{C}P^1$ are regular Lefshetz fibrations of 2-toruses. Now Corollary 3 follows from Corollary 2.

Theorem 10.*) Let $f: V \longrightarrow \Delta$ be an analytic fibration of elliptic curves, $\nu(f)$ be the number of multiple fibers of

*) This Theorem generalizes results of Kodaira's work on homotopy K3 surfaces (see [19]. The case $\nu(f) = 0$ was proved by A. Kas (see [18]).

$f: V \longrightarrow \Delta$. Then V is simply-connected if and only if the following conditions are satisfied:

(i) Δ is isomorphic to $\mathbb{C}P^1$,

(ii) there exists a fiber of $f: V \longrightarrow \Delta$ which (when reduced) is a singular curve;

(iii) $0 \leq \nu(f) \leq 2$ and in the case $\nu(f) = 2$ the multiplicities m_1, m_2 of multiple fibers are relatively prime numbers.

Proof. a) Suppose that $\pi_1(V) = 0$. Then (i) is evident. Proof of (iii) is contained in the proof of Proposition 2 of [19]. Consider (ii). Suppose that all fibers of $f: V \longrightarrow \Delta$ are non-singular curves. Let $a, b \in \Delta$ be such that for any $c \in \Delta - a - b$, $f^{-1}(c)$ is not a multiple fiber of $f: V \longrightarrow \Delta$, D_a and D_b be small closed 2-disks with the centers in a and b respectively, $s_a = \partial D_a$, $s_b = \partial D_b$. Since $f^{-1}(a)$ is non-singular, that is, a 2-torus, and $(f^{-1}(a).f^{-1}(a))_V = 0$ we have that the differential normal bundle of $f^{-1}(a)$ in V is trivial. Hence $f^{-1}(s_a)$ (corresp. $f^{-1}(D_a)$) is diffeomorphic to $s_a \times T^2$ (corresp. $D_a \times T^2$) ($f^{-1}(s_a)$ is the boundary of a regular neighborhood $f^{-1}(D_a)$ of $f^{-1}(a)$ in V). We obtain $\dim_{\mathbb{Q}} H_1(f^{-1}(s_a), \mathbb{Q}) = 3$. It is clear that $s_a \longrightarrow \overline{\Delta - D_a - D_b}$ is a homotopy equivalence. Hence $f^{-1}(s_a)$ is homotopy equivalent to $f^{-1}(\overline{\Delta - D_a - D_b})$ and $\dim_Q H_1(f^{-1}(\overline{\Delta - D_a - D_b}), \mathbb{Q}) = 3$. From $f^{-1}(D_a) \approx D_a \times T^2$ we easily get that

$$\dim_{\mathbb{C}} H_2(f^{-1}(D_a),\ f^{-1}(s_a);\mathbb{C}) = 1.$$

By the same reasons

$$\dim_{\mathbb{C}} H_2(f^{-1}(D_b),\ f^{-1}(s_b);\mathbb{C}) = 1.$$

It is clear that

$$H_2(V,f^{-1}(\overline{\Delta-D_a-D_b}),\mathbb{C}) = H_2(f^{-1}(D_a),f^{-1}(s_a);\mathbb{C})\oplus H_2(f^{-1}(D_b),f^{-1}(s_b));\mathbb{C}).$$

Hence $\dim_{\mathbb{C}} H_2(V,f^{-1}(\overline{\Delta-D_a-D_b});\mathbb{C}) = 2$ and from the exact sequence

$$H_2(V,f^{-1}(\overline{\Delta-D_a-D_b});\mathbb{C}) \longrightarrow H_1(f^{-1}(\overline{\Delta-D_a-D_b}),\mathbb{C}) \longrightarrow H_1(V,\mathbb{C})$$

we see that $H_1(V,\mathbb{C}) \neq 0$. Thus $\pi_1(V) \neq 0$. Contradiction. (ii) is proved.

b) Now suppose that conditions (i),(ii),(iii) are satisfied and prove $\pi_1(V) = 0$.

From Theorems 8 and 8a it follows that we can assume that all singular fibers of V are of type I_1 or ${}_mI_o$. Let $a,b \in \Delta$ be such that for any $c \in \Delta-a-b$, $f^{-1}(c)$ is not a multiple fibre of $f\colon V \longrightarrow \Delta$, $a_1,\cdots,a_\mu \in \Delta$ be such that $f^{-1}(a_i)$, $i = 1,\cdots,\mu$, are all the singular fibers of $f\colon V \longrightarrow \Delta$ which have type I_1. From Kodaira's theory of logarithmic transform it follows that there exists an analytic fibration of elliptic curves $\tilde{f}\colon \tilde{V} \longrightarrow \Delta$ such that $\tilde{f}|_{\tilde{f}^{-1}(\Delta-a-b)}\colon \tilde{f}^{-1}(\Delta-a-b) \longrightarrow \Delta-a-b$ is isomorphic to

$f|_{f^{-1}(\Delta-a-b)}: f^{-1}(\Delta-a-b) \longrightarrow \Delta-a-b$ and $\tilde{f}^{-1}(a), \tilde{f}^{-1}(b)$ are regular fibers of $\tilde{f}: \tilde{V} \longrightarrow \Delta$. Take $a_o \in \Delta-a-b - \bigcup_{i=1}^{\mu} a_i$. From Theorem 9 we obtain that there exists a set of disjoint smooth paths $\gamma_1, \cdots, \gamma_\mu$ in $\Delta-a-b$ connecting a_o with $a_1, \cdots, a_\mu$ such that the corresponding Lefshetz vanishing cycles $\delta_1, \cdots, \delta_\mu \in H_1(f^{-1}(a_o), \mathbb{Z})$ generate $H_1(f^{-1}(a_o), \mathbb{Z})$. We see that the image of $\pi_1(f^{-1}(a_o))$ in $\pi_1(V)$ is equal to zero.

Let D_a (corresp. D_b) be a small closed disk in $\Delta-b-\bigcup_{i=1}^{\mu} a_i$ (corresp. in $\Delta-a-\bigcup_{i=1}^{\mu} a_i$) with the center a (corresp. b), $s_a = \partial D_a$ (corresp. $s_b = \partial D_b$), $\tilde{s}_a$ (corresp. $\tilde{s}_b$) be a cross-section of $\tilde{f}|_{\tilde{f}^{-1}(\tilde{s}_a)}: \tilde{f}^{-1}(s_a) \longrightarrow s_a$ (corresp. $\tilde{f}|_{\tilde{f}^{-1}(s_b)}: \tilde{f}^{-1}(s_b) \longrightarrow s_b$).

Let $\bar{s}_a$ (corresp. $\bar{s}_b$) be a cross-section of $f|_{f^{-1}(s_a)}: f^{-1}(s_a) \rightarrow s_a$ (corresp. $f|_{f^{-1}(s_b)}: f^{-1}(s_b) \longrightarrow s_b$) which corresponds to $\tilde{s}_a$ (corresp. $\tilde{s}_b$) (recall that $\tilde{f}|_{\tilde{f}^{-1}(\Delta-a-b)}$ is isomorphic to $f|_{f^{-1}(\Delta-a-b)}$). Let m_a (corresp. m_b be the multiplicity of the fiber $f^{-1}(a)$ (corresp. $f^{-1}(b)$). (Possibly both or one of m_a, m_b are equal to one.) In [19], p. 68 (Proof of Lemma 6) Kodaira shows that we can choose $\tilde{s}_a$ (corresp. $\tilde{s}_b$) such that the loop $\bar{s}_a^{m_a}$ (corresp. $\bar{s}_b^{m_b}$) is homotopic on V to some loop in $f^{-1}(a_o)$. From $\mathrm{Im}[\pi_1(f^{-1}(a_o)) \longrightarrow \pi_1(V)] = 0$ we infer that $\bar{s}_a^{m_a}$ (corresp. $\bar{s}_b^{m_b}$) is homotopically equivalent to zero in V.

Let $\Delta' = \Delta-a-b-\bigcup_{i=1}^{\mu} a_i$, $f^{-1}(\Delta') = M'$, $f' = f|_{M'}: M' \longrightarrow \Delta'$, $C_o = f^{-1}(a_o)$.

Since $f'\colon M' \longrightarrow \Delta'$ is a fiber bundle, we have the following exact sequence:

$$(3) \qquad \pi_1(C_o) \xrightarrow{\beta'} \pi_1(M') \xrightarrow{\beta} \pi_1(\Delta').$$

Let D_i, $i = 1,2,\cdots,\mu$, be a small closed circle in $\Delta\text{-}a\text{-}b\text{-}\bigcup_{\substack{j=1\\ j\neq i}}^{\mu} a_j$ with the center a_i, $s_i = \partial D_i$. There exists a cross-section $\bar{s}_i$ of $f|_{f^{-1}(s_i)}\colon f^{-1}(s_i) \to s_i$. Using $\bar{s}_1,\cdots,\bar{s}_\mu,\bar{s}_a,\bar{s}_b$ we construct such elements $\sigma_1,\cdots,\sigma_\mu,\sigma_a,\sigma_b$ in $\pi_1(M')$ that $\beta(\sigma_1),\cdots,\beta(\sigma_\mu),\beta(\sigma_a),\beta(\sigma_b)$ generate $\pi_1(\Delta')$, $\beta(\sigma_a)\beta(\sigma_b^{-1})$ is in a subgroup of $\pi_1(M')$ generated by $\beta(\sigma_1),\cdots,\beta(\sigma_\mu)$, images of $\sigma_1,\cdots,\sigma_\mu$ in $\pi_1(V)$ are trivial and σ_a (corresp. σ_b) is a conjugate to the loop $\bar{s}_a$ (corresp. $\bar{s}_b$). It is clear that the canonical homomorphism $\varphi\colon \pi_1(M') \longrightarrow \pi_1(V)$ is an epimorphism. We see (from (3)) that any $z \in \pi_1(M')$ can be written in the form $z = z'\cdot z''$, where $z' \in \beta'(\pi_1(C_o))$, z'' is in the subgroup of $\pi_1(M')$ generated by $\sigma_1,\cdots,\sigma_\mu,\sigma_a$. Then $\varphi(z)$ is in the subgroup of $\pi_1(V)$ generated by $\varphi(\sigma_a)$ and $\varphi(\sigma_a)$ generates $\pi_1(V)$ (because $\pi_1(V) = \varphi(\pi_1(M'))$.

We have also that $\sigma_a\sigma_b^{-1}$ is in the subgroup of $\pi_1(M')$ generated by $\sigma_1,\cdots,\sigma_\mu$ and $\beta'(e_1),\beta'(e_2)$ where e_1,e_2 are generators of $\pi_1(C_o)$. We see that $\varphi(\sigma_a\sigma_b^{-1}) = 1$, that is, $\varphi(\sigma_a) = \varphi(\sigma_b)$. As we mentioned above, $\bar{s}_a^{m_a}$, $\bar{s}_b^{m_b}$ are homotopically equivalent to some loops in C_o. That means that $\sigma_a^{m_a}$, $\sigma_b^{m_b}$ are conjugate to some elements of $\beta'(\pi_1(C_o))$. From $\varphi(\beta'(\pi_1(C_o)) = 0$ and $\varphi(\sigma_a) = \varphi(\sigma_b)$ we obtain that $[\varphi(\sigma_a)]^{m_a} = [\varphi(\sigma_b)]^{m_b} = 0$. Because m_a and m_b are relatively prime we have $\varphi(\sigma_a) = 0$. Hence $\pi_1(V) = 0$. Q.E.D.

§3. Kodaira fibrations of 2-toruses.

We introduce now the following notations:

Let $k,m \in \mathbb{Z}$ be such that $m > 1$ and k is relatively prime to m. For $\epsilon > 0$ denote $D_\epsilon = \{\sigma \in \mathbb{C} \mid |\sigma| < \epsilon\}$. Let G be a group of automorphisms of $D_{\epsilon^{\frac{1}{m}}} \times \mathbb{C}$ consisting of transformations

$$(\sigma,\zeta) \rightarrow (\sigma,\zeta+n_1 i+n_2),\ n_1,n_2 \in \mathbb{Z}\,(\sigma \in D_{\epsilon^{\frac{1}{m}}},\ \zeta \in \mathbb{C})$$

and let $F(D_{\epsilon^{\frac{1}{m}}}) = D_{\epsilon^{\frac{1}{m}}} \times \mathbb{C}/G$. Denote by $[\sigma,\zeta]$ the point on $F(D_{\epsilon^{\frac{1}{m}}})$ corresponding to $(\sigma,\zeta) \in D_{\epsilon^{\frac{1}{m}}} \times \mathbb{C}$. Let $\mathcal{G}$ be a cyclic group of analytic automorphisms of $F(D_{\epsilon^{\frac{1}{m}}})$ generated by

$$g\colon [\sigma,\zeta] \longrightarrow [\rho\sigma,\ \zeta + \frac{k}{m}] \quad \text{where} \quad \rho = e^{\frac{2\pi i}{m}},$$

and $F_{m,k} = F(D_{\epsilon^{\frac{1}{m}}})/\mathcal{G}$. Denote by $[\sigma,\zeta]^{\sim}$ the point on $F_{m,k}$ corresponding to $[\sigma,\zeta] \in F(D_{\epsilon^{\frac{1}{m}}})$ and by $f_{m,k}\colon F_{m,k} \longrightarrow D_\epsilon$ the map given by $f_{m,k}([\sigma,\zeta]^{\sim}) = \sigma^m$.

<u>Definition 8</u>. Let $f\colon M \longrightarrow S$ be a differential map of compact oriented differential manifolds, $\dim M = 4$, $\dim S = 2$, $\partial S = \emptyset$. We say that $f\colon M \longrightarrow S$ is a Kodaira fibration with ν multiple fibers where $\nu \in \mathbb{Z}$, $\nu \geq 0$ if the following is true:

a) if $\nu = 0$, $f: M \longrightarrow S$ is a Lefshetz fibration of 2-toruses,

b) if $\nu > 0$ then there exist ν points, say $c_1, \cdots, c_\nu \in S$, ν closed disjoint 2-disks $E_1, \cdots, E_\nu \subset S$ with the centers in $c_1, \cdots, c_\nu$ respectively, ν pairs of integers $(m_1, k_1), \cdots, (m_\nu, k_\nu)$, where for any $j = 1, 2, \cdots, \nu$, $m_j > 1$ and k_j is relatively prime to m_j, and ν commutative diagrams

$$\begin{array}{ccc} f^{-1}(E_j) & \xrightarrow{\psi_j} & F_{m_j, k_j} \\ {\scriptstyle f|f^{-1}(E_j)} \downarrow & & \downarrow {\scriptstyle f_{m_j, k_j}} \\ E_j & \xrightarrow[\varphi_j]{} & D_\epsilon \end{array} \quad , \qquad j = 1, 2, \cdots, \nu,$$

such that

(i) $$f\Big|_{M - \bigcup_{j=1}^{\nu} f^{-1}(E_j)} : M - \bigcup_{j=1}^{\nu} f^{-1}(E_j) \longrightarrow S - \bigcup_{j=1}^{\nu} E_j$$

is a Lefshetz fibration of 2-toruses;

(ii) for any $j = 1, 2, \cdots, \nu$, φ_j and ψ_j are orientation preserving diffeomorphisms (where orientations of D_ϵ and F_{m_j, k_j} are defined by complex structure).

Let $T(f)$ $(\subset S)$ be the set of critical values of $f: M \longrightarrow S$ if $\nu = 0$ and the set of critical values of

$$f\Big|_{M-\bigcup_{j=1}^{\nu} f^{-1}(E_j)} : M-\bigcup_{j=1}^{\nu} f^{-1}(E_j) \longrightarrow S - \bigcup_{j=1}^{\nu} E_j$$

if $\nu > 0$, $T'(f) = \emptyset$ if $\nu = 0$ and $T'(f) = \{c_1,\cdots,c_\nu\}$ if $\nu > 0$. We call $T(f)$ (corresp. $T'(f)$ set of non-degenerate (corresp. degenerate) critical values of $f\colon M \to S$.

Definition 8a. Let $f\colon M \longrightarrow S$ be a Kodaira fibration with ν multiple fibers and in the case $\nu > 0$ let $\{c_1,\cdots,c_\nu\}$ be the set of degenerate critical values of $f\colon M \longrightarrow S$, and $\{E_1,\cdots,E_\nu\}$, $\{(m_1,k_1),\cdots,(m_\nu,k_\nu)\}$, $\{(\varphi_1,\psi_1),\cdots,(\varphi_\nu,\psi_\nu)\}$ be the same as in Definition 8. Let $D_\epsilon^{\cdot} = \{\sigma \in \mathbb{C} \,|\, 0 < |\sigma| < \epsilon\}$,

$$F^{\cdot}(D_\epsilon) = \{[\sigma,\zeta] \in F(D_\epsilon) \,|\, \sigma \neq 0\},\ F^{\cdot}_{m,k} = f^{-1}_{m,k}(D^{\cdot}_\epsilon).$$

Following Kodaira we define a map

$$\Lambda_{m,k}\colon F^{\cdot}_{m,k} \longrightarrow F^{\cdot}(D_\epsilon)$$

by $\Lambda_{m,k}([\sigma,\zeta]^{\sim}) = [\sigma^m,\ \zeta - \frac{k}{2\pi i}\log\sigma]$.

Let $\nu > 0$ and $F(D_\epsilon)_j$ (corresp. $D_{\epsilon,j}$), $j = 1,2,\cdots,\nu$, be ν copies of $F(D_\epsilon)$ (corresp. D_ϵ).

Define a new 4-manifold $\tilde{M}$ as union of $M - \bigcup_{j=1}^{\nu} f^{-1}(c_j)$ and $F(D_\epsilon)_1,\cdots,F(D_\epsilon)_\nu$ where we identify $x \in f^{-1}(E_j - c_j)$, $j = 1,2,\cdots,\nu$, with $\Lambda_{m_j,k_j}\psi_j(x) \in F^{\cdot}(D_\epsilon)_j$.

Define the 2-manifold $\tilde{S}$ as union of $S - \bigcup_{j=1}^{\nu} c_j$ and $D_{\epsilon,1},\cdots,D_{\epsilon,\nu}$ where we identify $a \in E_j - c_j$, $j = 1,2,\cdots,\nu$, with $\varphi_j(a) \in D_{\epsilon,j}$. Let $\tilde{f}\colon \tilde{M} \longrightarrow \tilde{S}$ be a map defined by

$$\tilde{f}(x) = \begin{cases} f(x) & \text{if} \quad x \in M - \bigcup_{j=1}^{\nu} c_j \\ \pi_j(x) & \text{if} \quad x \in F(D_\epsilon)_j \text{ where } \pi_j([\sigma,\zeta]) = \sigma. \end{cases}$$

In the case $\nu = 0$ we take $\tilde{M} = M$, $\tilde{S} = S$, $\tilde{f} = f$. It is clear that $\tilde{f}\colon \tilde{M} \longrightarrow \tilde{S}$ is a Lefshetz fibration of 2-toruses and that the set of non-degenerate critical values of $f\colon M \longrightarrow S$ coincides with the set of critical values of $\tilde{f}\colon \tilde{M} \longrightarrow \tilde{S}$ (by evident embedding $S - \bigcup_{j=1}^{\nu} c_j \longrightarrow \tilde{S}$).

We shall call $\tilde{f}\colon \tilde{M} \longrightarrow \tilde{S}$ the Lefshetz fibration corresponding to Kodaira fibration $f\colon M \longrightarrow S$.

Definition 9. Let $f\colon M \longrightarrow S$ be a Kodaira fibration. We call $f\colon M \longrightarrow S$ regular if S is diffeomorphic to a 2-sphere and the set of non-degenerate critical values of f is not empty.

Let $e(M)$ be the Euler characteristic of M. We define $\chi(M) = \frac{e(M)}{12}$ and (following F. Hirzebruch) call it arithmetical genus of M.

Lemma 9. Let $f: M \longrightarrow S$ be a Kodaira fibration with ν multiple fibers, $\tilde{f}: \tilde{M} \longrightarrow \tilde{S}$ be the corresponding Lefshetz fibration.

Then i) $e(M) = e(\tilde{M})$;

ii) if $f: M \longrightarrow S$ is regular, then $e(M) > 0$ and $e(M) \equiv 0 \pmod{12}$, that is, $\chi(M)$ is a positive integer.

Proof. i) Let $\{c_1, \cdots, c_\nu\}$ be the set of degenerate critical values of $f: M \longrightarrow S$. It is clear that $e(f^{-1}(c_j)) = 0$, $e(F(D_\epsilon)_j - F^{\cdot}(D_\epsilon)_j) = 0$, $j = 1, 2, \cdots, \nu$. We have

$$e(M) = e(M, \bigcup_{j=1}^{\nu} f^{-1}(c_j)), \qquad e(\tilde{M}) = e(M, \bigcup_{j=1}^{\nu} \tilde{f}^{-1}(c_j)).$$

Hence $e(M) = e(\tilde{M})$.

ii) Immediately follows from i) and Theorem 9. ($\tilde{f}: \tilde{M} \longrightarrow \tilde{S}$ is a regular Lefshetz fibration if $f: M \longrightarrow S$ is regular).

Q.E.D.

Lemma 10. Let $f: M \longrightarrow S$ be a regular Kodaira fibration with ν multiple fibers. Suppose that $\nu \leq 1$ and $\chi(M) = 1$. Then M is diffeomorphic to $P \# 9Q$.

Proof. If $\nu = 0$ then Lemma 10 follows from Theorem 9. Consider the case $\nu = 1$. Let $\tilde{f}: \tilde{M} \longrightarrow \tilde{S}$ be the Lefshetz fibration corresponding to $f: M \longrightarrow S$. By Lemma 9 $\chi(\tilde{M}) = \chi(M) = 1$

and by Theorem 9 $\tilde{f}: \tilde{M} \longrightarrow \tilde{S}$ is isomorphic to $f_o: M_o \longrightarrow S_o$. Identify $\tilde{f}: \tilde{M} \longrightarrow \tilde{S}$ with $f_o: M_o \longrightarrow S_o$. Let c_1 be the degenerate critical value of $f: M \longrightarrow S$ and E_1, $(m_1,k_1),(\varphi_1,\psi_1)$ be the same as in Definition 8. Identify $M-f^{-1}(c_1), F(D_\epsilon)_1$ (corresp. $S-c_1, D_{\epsilon,1}$) with their images in $\tilde{M}$ (corresp. $\tilde{S}$). Let $\tilde{c}_1$ be the center of $D_{\epsilon,1}$. Without loss of generality we can assume that $D_{\epsilon,1}$ (considered in S_o) is contained in some coordinate neighborhood U of $\tilde{c}_1$ in S_o ($= \mathbb{C}P^1$) with complex coordinate τ and that $\tau(\tilde{c}_1) = 0$ and $D_{\epsilon,1}$ is given in U by $|\tau| < \epsilon$. Let e_{1a}, e_{2a}, $a \in D_{\epsilon,1}$, be a basis of $H_1(\tilde{f}^{-1}(a),\mathbb{Z})$ corresponding to the vectors $i,1 \in \mathbb{C}$ by our identification $\tilde{f}^{-1}(D_{\epsilon,1}) = F(D_\epsilon)_1 = D_{\epsilon,1} \times \mathbb{C}/G$. Using the family $\{e_{1a}, e_{2a}; a \in D_{\epsilon,1}\}$ we can identify the complex manifold $f_o^{-1}(D_{\epsilon,1})$ with $D_{\epsilon,1} \times \mathbb{C}/G_\omega$ where G_ω is the group consisting of analytic automorphisms

$$(\tau,\zeta) \longrightarrow (\tau, \zeta + n_1\omega(\tau) + n_2),\ n_1,n_2 \in \mathbb{Z}\ \ (\tau \in D_{\epsilon,1},\ \zeta \in \mathbb{C})$$

and $\omega(\tau)$ is a holomorphic function in $D_{\epsilon,1}$ with $\operatorname{Im}\omega(\tau) > 0$. Let $\epsilon^* = \epsilon^{\frac{1}{m_1}}$, $D_{\epsilon^*} = \{\sigma \in \mathbb{C} \mid |\sigma| < \epsilon^*\}$, G^*_ω be the group of analytic automorphisms of $D_{\epsilon^*} \times \mathbb{C}$ consisting of transformations

$$(\sigma,\tau) \longrightarrow (\sigma, \zeta + n_1\omega(\sigma^{m_1}) + n_2),\ n_1,n_2 \in \mathbb{Z}\ (\sigma \in D_{\epsilon^*},\ \zeta \in \mathbb{C})$$

and $F(D_{\epsilon^*},\omega) = D_{\epsilon^*} \times \mathbb{C}/G^*_\omega$. As above we denote by $[\sigma,\zeta]$ the point on $F(D_{\epsilon*},\omega)$ corresponding to $(\sigma,\zeta) \in D_{\epsilon*}\times\mathbb{C})$. Let $\mathcal{G}_\omega$ be the cyclic group of analytic automorphisms of $F(D_{\epsilon*},\omega)$ generated by the transformation

$$[\sigma,\zeta] \longrightarrow [\rho\sigma,\ \zeta + \frac{k_1}{m_1}],\ \rho = e^{\frac{2\pi i}{m_1}}$$

and $F_{m_1,k_1,\omega} = F(D_{\epsilon*},\omega)/\mathcal{G}_\omega$. Denote by $[\sigma,\zeta]^\sim$ the point on F_{m_1,k_1,ω_1} corresponding to $[\sigma,\zeta] \in F(D_{\epsilon*},\omega)$. Let

$$D^{\cdot}_{\epsilon,1} = \{\tau \in D_{\epsilon,1} \,|\, \tau \neq 0\},$$

$$F^{\cdot}_{m_1,k_1,\omega} = \{[\sigma,\zeta]^\sim \in F_{m_1,k_1,\omega_1} \,|\, \sigma \neq 0\}.$$

Define a holomorphic map

$$\Lambda_{m_1,k_1,\omega_1} : F^{\cdot}_{m_1,k_1,\omega} \longrightarrow f_o^{-1}(D^{\cdot}_{\epsilon,1}) \quad \text{by}$$

$$\Lambda_{m_1,k_1,\omega_1}([\sigma,\zeta]^\sim) = [\sigma^{m_1},\ \zeta - \frac{k_1}{2\pi i}\log\sigma]_{M_o,\omega}$$

where we denote by $[\tau,\zeta]_{M_o,\omega}$ the point on $f_o^{-1}(D_{\epsilon,1})$ corresponding to $(\tau,\zeta) \in D_{\epsilon,1}\times\mathbb{C}$ by our identification $f_o^{-1}(D_{\epsilon,1}) = D_{\epsilon,1}\times\mathbb{C}/G_\omega$. Define a new complex manifold $\hat{M}_o$ as union of $F_{m_1,k_1,\omega}$ and $M_o - f_o^{-1}(\tilde{C}_1)$ where we identify $x \in F^{\cdot}_{m_1,k_1,\omega} (\subset F_{m_1,k_1,\omega})$ with

$$\Lambda_{m_1,k_1,\omega}(x) \in f_o^{-1}(D^{\cdot}_{\epsilon,1}) (\subset M_o - f_o^{-1}(\tilde{C}_1)).$$

It is clear that $\hat{M}_o$ is obtained from M_o by Kodaira's logarithmic transform at $\tilde{c}_1$ (see [14], p. 768). Let $\hat{f}_o: \hat{M}_o \longrightarrow S_o$ be the holomorphic map canonically corresponding to $f_o: M_o \longrightarrow S_o$.

Let K_{M_o} (corresp. $K_{\hat{M}_o}$) be the canonical bundle of M_o (corresp. $\hat{M}_o$). Using $K_{M_o} = -[f_o^{-1}(a)]$, $a \in S_o - \tilde{c}_1$ and the Kodaira formula for canonical class for elliptic surface with multiple fibers (see [14], p. 772) we have

$$K_{\hat{M}_o} = -[f_o^{-1}(a)] + (m_1-1)[f_o^{-1}(\tilde{c}_1)].$$

Evidently $[f_o^{-1}(a)] = m_1[f_o^{-1}(\tilde{c}_1)]$. Hence $K_{\hat{M}_o} = -[f_o^{-1}(\tilde{c}_1)]$ and all puri-genuses of $\hat{M}_o$ vanish. Since $\pi_1(\hat{M}_o) = 0$ (Theorem 10) we have that $\hat{M}_o$ is a rational surface. Since $e(\hat{M}_o) = e(M_o) = 12$ we have $b_2(\hat{M}_o) = 10$. $\hat{M}_o$ could be obtained from some minimal rational surface by σ-processes. Because any minimal rational surface is diffeomorphic either to P or to $P \# Q$ or to $S^2 \times S^2$ we see that $\hat{M}_o$ is diffeomorphic to $P \# 9Q$.

We shall prove now that M is diffeomorphic to $\hat{M}_o$.

Let $u_1(\tau) = \operatorname{Re} \omega(\tau)$, $v_1(\tau) = \operatorname{Im} \omega(\tau)$ and $A_\tau: \mathbb{C} \longrightarrow \mathbb{C}$ be a diffeomorphism defined by

$$\zeta \longrightarrow \operatorname{Re} \zeta + u_1(\tau) \operatorname{Im} \zeta + i v_1(\tau) \operatorname{Im} \zeta$$

(A_τ is a non-degenerate real-linear orientation-preserving transformation because $v_1(\tau) = \operatorname{Im} \omega(\tau) > 0$). As above, we use for the points of $f^{-1}(E_1)$ (corresp. $\hat{f}_o^{-1}(D_{\epsilon,1})$) the notation $[\sigma,\zeta]_{\tilde{M}}$ (corresp. $[\sigma,\zeta]_{\tilde{\hat{M}}_o}$). Define a map

$$A\colon f^{-1}(E_1) \longrightarrow \hat{f}_o^{-1}(D_{\epsilon,1}) \text{ by}$$
$$[\sigma,\zeta]_{\tilde{M}} \longrightarrow [\sigma, A_\sigma m_1(\zeta)]_{\tilde{\hat{M}}_o} .$$

It is easy to verify that A is a diffeomorphism and that

$$(\hat{f}_o|_{\hat{f}_o^{-1}(D_{\epsilon,1})})\cdot A = f|_{f^{-1}(E_1)} .$$

Let $\epsilon' = \frac{\epsilon}{2}$, $\bar{D} = \{\tau \in D_{\epsilon,1} \,\big|\, |\tau| \le \epsilon'\}$, $\bar{E} = \varphi_1^{-1}(\bar{D})$, $\underline{s} = \partial\bar{D}$.

$\bar{A} = A|_{f^{-1}(\partial\bar{E})}\colon\ f^{-1}(\partial\bar{E}) \longrightarrow \hat{f}_o^{-1}(\partial\bar{D})$, $S' = \overline{S-\bar{E}}$, $S_o' = \overline{S_o-\bar{D}}$,

$M' = f^{-1}(S')$, $M_o' = f_o^{-1}(S_o')$, $\hat{M}_o' = f_o^{-1}(S_o')$.

Let $i_M\colon M' \longrightarrow M_o'$, $i_{\hat{M}_o}\colon \hat{M}_o' \longrightarrow M_o'$ be isomorphisms corresponding to our constructions of $\tilde{f}\colon \tilde{M} \longrightarrow \tilde{S}$ and $\hat{f}_o\colon \hat{M}_o \longrightarrow S_o$, $\underline{\tilde{\alpha}}\colon f_o^{-1}(\underline{s}) \longrightarrow f_o^{-1}(\underline{s})$ be a diffeomorphism which is equal to

$$\left[i_{\hat{M}_o}|_{\hat{f}_o^{-1}(\underline{s})}\right] \cdot \bar{A} \cdot \left[(i_M^{-1})|_{f_o^{-1}(\underline{s})}\right] .$$

Let $\underline{\alpha}$ be an element in $\Omega(T^2)$ corresponding to $\underline{\tilde{\alpha}}$ and the trivialization of $f_o^{-1}(\underline{s}) \longrightarrow \underline{s}$ given by $f_o^{-1}(\underline{s}) \subset f_o^{-1}(D_{\epsilon,1}) = D_{\epsilon,1} \times \mathbb{C}/G$. Using our choice of $\{e_{1,a}, e_{2,a}; a \in D_{\epsilon,1}\}$ for the identification $f_o^{-1}(D_{\epsilon,1}) = D_{\epsilon,1} \times \mathbb{C}/G_\omega$ it is easy to verify that $\underline{\alpha} \in \Omega_o(T^2)$. Now by Remark to Lemma 7 (see p. 168) we obtain a diffeomorphism $\tilde{\beta}|_{f_o^{-1}(\underline{s})} = \underline{\tilde{\alpha}}$. Define $\tilde{A}: M \longrightarrow \hat{M}_o$ by

$\tilde{A}|_{f^{-1}(\bar{E})} = A|_{f^{-1}(\bar{E})}$, $\tilde{A} \mid_{f^{-1}(S')} = (i_{\hat{M}_o})^{-1} \cdot \beta \cdot i_M$. We see that M is diffeomorphic to $\hat{M}_o$. Thus M is diffeomorphic to $P \# 9Q$. Q.E.D.

Definition 10. Let $f_i: M_i \to S_i$, $i = 1,2$, be two Kodaira fibrations. We define direct sum of Kodaira fibrations $f_1 \oplus f_2: M_1 \oplus M_2 \longrightarrow S_1 \# S_2$ by the same way that in Definition 7 we defined direct sum of Lefshetz fibrations (see Def. 7, p. 174).

Lemma 11. Let $f_i: M_i \to S_i$, $i = 1,2$, be two Kodaira fibrations, $f: M \to S$ be a Kodaira fibration which is isomorphic to $f_1 \oplus f_2: M_1 \oplus M_2 \longrightarrow S_1 \# S_2$. Suppose that $\pi_1(M_1) = 0 = \pi_1(M) = 0$ and the intersection form of M_1 is of odd type.

Let U be an open 2-disk in S_2 which does not contain critical values of f_2, $b,c \in U$, γ_{bc} be a smooth path in U connecting b with c, $s_b \subset f_2^{-1}(b)$, $s_c \subset f_2^{-1}(c)$ be two smooth circles such that $\gamma_{bc*} s_b$, s_c generate $H_1(f^{-1}(c), \mathbb{Z})$ where

$\gamma_{bc*}: H_1(f^{-1}(b),\mathbb{Z}) \longrightarrow H_1(f^{-1}(c),\mathbb{Z})$ is the canonical isomorphism corresponding to γ_{bc}. Then

i) $M \# P \# Q$ is diffeomorphic to $M_1 \# \tilde{\tilde{M}}_2$, where $\tilde{\tilde{M}}_2$ is a 4-manifold obtained from M_2 by surgeries along s_b and s_c;

ii) if $f_1: M_1 \longrightarrow S_1$ is isomorphic to $f_o: M_o \longrightarrow S_o$ (see the formulation of Theorem 9) then $M \# P$ is diffeomorphic to

$$P \# 8Q \# \tilde{\tilde{M}}_2 .$$

Proof. Let $a^{(i)} \in S_i$, $i = 1,2$ be some non-critical value of f_i, $C_{(i)} = f_i^{-1}(a^{(i)})$, $\pi_{(i)}: TC_{(i)} \longrightarrow C_{(i)}$ be a tubular neighborhood of $C_{(i)}$ in M_i. Since $(C^2_{(i)})_{M_i} = 0$, $i = 1,2$, there exists an isomorphism $\tilde{\beta}$ of fiber bundles $\pi_{(1)}|\partial TC_{(1)}: \partial TC_{(1)} \longrightarrow C_{(1)}$ and $\pi_{(2)}|\partial TC_{(2)}: \partial TC_{(2)} \longrightarrow C_{(2)}$ which reverses orientation of fiber. Let $\beta: C_{(1)} \longrightarrow C_{(2)}$ be an isomorphism of bases corresponding to $\tilde{\beta}$. Using the definition of direct sum of Kodaira fibrations, we see that we can identify M with $[M_1 - TC_{(1)}] \cup_{\tilde{\beta}} [M_2 - TC_{(2)}]$. Let $x_1 \in C_{(1)}$, E_1 be a small open 2-disk on $C_{(1)}$ with the center x_1, $T\dot{C}_{(1)} = \pi^{-1}_{(1)}(C_{(1)} - E_1)$, $T\dot{C}_{(2)} = \pi^{-1}_{(2)}(C_{(2)} - \beta E_1)$. Consider $s = \partial(\pi^{-1}_{(1)}(x_1))$ as a subset in M. Let $M^{\sim}$ be a 4-manifold obtained from M by surgery along s. Now using the same arguments as on pages 45-47 we see that there exists an orientation reversing diffeomorphism $\beta^{\cdot}: \partial TC^{\cdot}_{(1)} \longrightarrow \partial TC^{\cdot}_{(2)}$ such that $M^{\sim}$ is diffeomorphic to

$$\overline{[M_1 - TC^{\cdot}_{(1)}]} \cup_{\beta^{\cdot}} \overline{[M_2 - TC^{\cdot}_{(2)}]}.$$

Now suppose that $f_1: M_1 \longrightarrow S_1$ is isomorphic to $f_o: M_o \longrightarrow S_o$. We know that M_o is isomorphic to $\mathbb{C}P^2$ with nine σ-processes such that the corresponding exceptional curves, say $\mathcal{E}^{(1)}, \dots, \mathcal{E}^{(9)}$, are cross-sections of $f_o: M_o \longrightarrow S_o$. Identifying $f_1: M_1 \longrightarrow S_1$ with $f_o: M_o \longrightarrow S_o$ we can assume that $x_1 = \mathcal{E}^{(1)} \cap C_{(1)}$ and $\pi^{-1}_{(1)}(x_1) = \mathcal{E}^{(1)} \cap TC_{(1)}$. Let $\sigma_1: M_1 \longrightarrow \underline{M}_1$ be a canonical contraction of $\mathcal{E}^{(1)}$ to a point, $\underline{C}^{\cdot}_{(1)} = \sigma_1 C^{\cdot}_{(1)}$, $T\underline{C}^{\cdot}_{(1)} = \sigma(TC^{\cdot}_{(1)})$. It is evident that we can identify $TC^{\cdot}_{(1)}$ with $T\underline{C}^{\cdot}_{(1)}$ (by identification $M_1 - \mathcal{E}^{(1)}$ with $\underline{M}_1 - \sigma_1(\mathcal{E}^{(1)})$. Using the same arguments as on page 48 we see that $M \# P$ is diffeomorphic to

$$[\underline{M}_1 - T\underline{C}^{\cdot}_{(1)}] \cup_{\beta^{\cdot}} [M_2 - TC^{\cdot}_{(2)}].$$

Now applying a result of R. Mandelbaum (see [5]) which we mentioned and used on page 55 we get that $M^{\sim}$ is diffeomorphic to $M_1 \# \tilde{\tilde{M}}_2$ (where $\tilde{\tilde{M}}_2$ is defined as in the formulation of the Lemma). In the case when $f_1: M_1 \longrightarrow S_1$ is isomorphic to $f_o: M_o \longrightarrow S_o$ we obtain that $M \# P$ is diffeomorphic to $\underline{M}_1 \# \tilde{\tilde{M}}_2$, that is, to $P \# 8Q \# \tilde{\tilde{M}}_2$. Since the intersection form of

M_1 is of odd type we have that the intersection form of $M^{\sim}$ is also of odd type. Using $\pi_1(M) = 0$ and results of Wall (see [8]) we obtain that $M^{\sim}$ is diffeomorphic to $M \# P \# Q$. Thus $M \# P \# Q \approx M_1 \# \tilde{\tilde{M}}_2$ and in case ii) $M \# P \approx P \# 8Q \# \tilde{\tilde{M}}_2$. Q.E.D.

Lemma 12. Let $f: M \rightarrow S$ be a Kodaira fibration with ν multiple fibers. Then M is simply-connected if and only if the following conditions are satisfied:

(i) $f: M \rightarrow S$ is regular;

(ii) $0 \leq \nu \leq 2$ and in the case $\nu = 2$ the corresponding multiplicities m_1 and m_2 are relatively prime.

Proof. That is almost word-by-word repetition of the proof of Theorem 10. Q.E.D.

Lemma 13. Let L be a 3-dimensional manifold diffeomorphic to a lense space (see [20]), $M = L \times S^1$, $C_1 = p \times S^1$ where $p \in L$, C_2 be a smooth circle in M with $C_2 \cap C_1 = \emptyset$, N be a 4-manifold obtained from M by surgeries along C_1 and C_2. Suppose that $\pi_1(N) = 0$. Then N is diffeomorphic to an S^2-bundle over S^2 (that is, to $S^2 \times S^2$ or to $P \# Q$).

Proof. Let $\tilde{S}_i = \{\tilde{\tau}_i \in \mathbb{C} \,|\, |\tilde{\tau}_i| = 1\}$, $S_i = \{\tau_i \in \mathbb{C} \,|\, |\tau_i| = 1\}$, $i = 1,2,3$, $\tilde{D}_1 = \{\tilde{\tau}_1 \in \mathbb{C} \,|\, |\tilde{\tau}_1| \leq 1\}$, $D_1 = \{\tau_1 \in \mathbb{C} \,|\, |\tau_1| \leq 1\}$,

$\tilde{Z} = \tilde{D}_1 \times \tilde{S}_2$, $Z = D_1 \times S_2$, $\tilde{Y} = \tilde{Z} \times \tilde{S}_3$, $Y = Z \times S_3$. We can identify L with $\tilde{Z} \cup_\varphi Z$ where $\varphi: \partial\tilde{Z} \longrightarrow \partial Z$ is defined by

$\varphi(\tilde{\tau}_1, \tilde{\tau}_2) = (\tilde{\tau}_1^a \tilde{\tau}_2^b, \tilde{\tau}_1^c \tilde{\tau}_2^d)$, $a,b,c,d \in \mathbb{Z}$, $\det \left\| \begin{matrix} a & b \\ c & d \end{matrix} \right\| = 1$. Now we can assume that $M = \tilde{Y} \cup_\psi Y$, $\psi: \partial\tilde{Y} \longrightarrow \partial Y$ is defined by $\psi(\tilde{\tau}_1, \tilde{\tau}_2, \tilde{\tau}_3) = \varphi(\tilde{\tau}_1, \tilde{\tau}_2) \times \tilde{\tau}_3$, and $p \in Z$, $p = (0,1)$, $C_1 = p \times S^3$. Denote by

$I_o = \{x \in S_2, -\frac{\pi}{4} \leq \arg \tau_2(x) \leq \frac{\pi}{4}\}$, $I'_o = \overline{S_2 - I_o}$,

$D(\tau_3) = D_1 \times I_o \times y$, $y \in S_3$, $\tau_3(y) = \tau_3$, $TC_1 = \bigcup_{|\tau_3| = 1} D(\tau_3)$ (union in Y).

Let $\lambda: TC_1 \longrightarrow C_1$ be defined by

$$\lambda(\tau_1, \tau_2, \tau_3) = (0, 1, \tau_3).$$

We can consider $\lambda: TC_1 \longrightarrow C_1$ as a tubular neighborhood of C_1 in M. Now we have two possibilities for a surgery of M along C_1 which correspond to two non-equivalent trivializations of $\lambda: TC_1 \longrightarrow C_1$. We can assume that these two trivializations, say $f_o: TC_1 \longrightarrow D(1) \times S_3$, $f_1: TC_1 \longrightarrow D(1) \times S_3$, are defined as follows:

$$f_o(\tau_1, \tau_2, \tau_3) = ((\tau_1, \tau_2, 1), \tau_3), \quad f_1(\tau_1, \tau_2, \tau_3) = ((\tau_1 \tau_3^{-1}, \tau_2, 1), \tau_3).$$

Define an autodiffeomorphism $\alpha: Y \longrightarrow Y$ by $\alpha(\tau_1, \tau_2, \tau_3) = (\tau_1 \tau_3, \tau_2, \tau_3)$. Let $\alpha' = \alpha|_{\partial Y}: \partial Y \longrightarrow \partial Y$ and $\tilde{\alpha}' = \psi^{-1} \alpha \psi: \partial\tilde{Y} \longrightarrow \partial\tilde{Y}$. Let $\left\| \begin{matrix} a' & b' \\ c' & d' \end{matrix} \right\| = \left\| \begin{matrix} a & b \\ c & d \end{matrix} \right\|^{-1}$. It is easy to verify that $\tilde{\alpha}'(\tilde{\tau}_1, \tilde{\tau}_2, \tilde{\tau}_3) = (\tilde{\tau}_1 \tilde{\tau}_3^{a'}, \tilde{\tau}_2 \tilde{\tau}_3^{c'}, \tilde{\tau}_3)$. Now define an autodiffeomorphism $\widetilde{\alpha}: \widetilde{Y} \longrightarrow \widetilde{Y}$ by $\widetilde{\alpha}(\widetilde{\tau}_1, \widetilde{\tau}_2, \widetilde{\tau}_3) = (\widetilde{\tau}_1 \widetilde{\tau}_3^{a'}, \widetilde{\tau}_2 \widetilde{\tau}_3^{c'}, \widetilde{\tau}_3)$. We obtain an isomorphism $\beta: M \longrightarrow M$ with $\beta|_{\widetilde{Y}} = \widetilde{\alpha}$, $\beta_{\downarrow Y} = \alpha$. It is easy to see that $\beta(C_1) = C_1$, $\beta(TC_1) = TC_1$, $\lambda\beta = \lambda$. Since

$$(\beta^{-1}|_{D(1)} \times id) \cdot f_1 \cdot \beta(\tau_1, \tau_2, \tau_3) = (\beta^{-1}|_{D(1)} \times id) \cdot f_1(\tau_1\tau_3, \tau_2, \tau_3) =$$

$$= (\beta^{-1}|_{D(1)} \times id)((\tau_1, \tau_2, 1), \tau_3) = ((\tau_1, \tau_2, 1), \tau_3)$$ we have that

$(\beta^{-1}|_{D(1)} \times id) \cdot f_1 \cdot \beta = f_o$. Hence β^{-1} transforms f_1 in f_o and we have to consider only the case when our surgery corresponds to the trivialization f_o. Denote by $M[C_1]$ a 4-manifold obtained from M by surgery along C_1 corresponding to f_o. We can construct $M[C_1]$ as follows: Let $D_3 = \{\tau_3 \in \mathbb{C} \,|\, |\tau_3| \leq 1\}$. Then $M[C_1] = \overline{M-T(C_1)} \cup_{f'_o} (\partial D(1) \times D_3)$, where

$$f'_o = f_o|_{\partial T(C_1)}: \partial T(C_1) \longrightarrow \partial D(1) \times S_3.$$

Now for any $x \in D_1$ denote by

$$S^2(x) = (x \times I'_o \times S_3) \cup (x \times \partial I'_o \times D_3)$$

where union is taken in

$$M[C_1]((x \times I'_o \times S_3 \in \overline{M-T(C_1)}, (x \times \partial I'_o \times D_3) \in \partial D(1) \times D_3).$$

Let

$$x_o \in D_1, \ \tau_1(x_o) = 0, \ TS^2(x_o) = \bigcup_{x \in D_1} S^2(x) \text{ (union in } M[C_1]) \text{ and}$$

$$\lambda': TS^2(x_o) \longrightarrow S^2(x_o)$$

be defined as follows: If $z \in S^2(x)$, $x \in D_1$, $z = x\times y$, where

$$y \in ((I_o' \times S_3) \cup (\partial I_o' \times D_3)$$

(boundaries are identified by evident way), then $\lambda'(z) = x_o \times y$.

Identify $S^2(x_o)$ with $(I_o' \times S_3) \cup (\partial I_o' \times D_3)$ and let $D^3 = I_o' \times D_3$ be a 3-disk with boundary $(I_o' \times S_3) \cup (\partial I_o' \times D_3) = S^2(x_o)$. Now identify $\lambda' \colon TS^2(x_o) \longrightarrow S^2(x_o)$ with $pr \colon D_1 \times S^2(x_o) \longrightarrow S^2(x_o)$ and let X be a 4-manifold obtained from $M[C_1]$ by surgery along $S^2(x_o)$ corresponding to the given trivialization. We have

$$X = \overline{M[C_1]} - \overline{TS^2(x_o)} \cup (S_1 \times I_o' \times D_3)$$

where the boundaries are identified by evident way.

Using

$$\overline{M-TC_1} = \overline{M-(D_1 \times I_o \times S_3)} = (\tilde{Z} \times \tilde{S}_3) \cup (D_1 \times I_o' \times S_3)$$

and

$$\overline{M[C_1]-TS^2(x_o)} = \overline{\{[(\tilde{Z} \times \tilde{S}_3) \cup (D_1 \times I_o' \times S_3)] \cup [\partial D(1) \times D_3]\} -}$$

$$\overline{- \{(D_1 \times I_o' \times S_3) \cup (D_1 \times \partial I_o \times D_3)\}} = (\tilde{Z} \times \tilde{S}_3) \cup (S_1 \times I_o \times D_3)$$

we have

$$X = [(\tilde{Z} \times \tilde{S}_3) \cup (S_1 \times I_o \times D_3)] \cup [S_1 \times I_o' \times D_3].$$

We see that we can identify X with

$$(\tilde{Z} \times \tilde{S}_3) \cup (S_1 \times S_2 \times D_3) = (\tilde{Z} \times \tilde{S}_3) \cup_{\psi} (\partial Z \times D_3)$$

which is isomorphic to $(\widetilde{Z}\times\widetilde{S}_3)\cup_{id}(\partial\widetilde{Z}\times\widetilde{D}_3)$, where $\widetilde{D}_3 = \{\widetilde{\tau}_3 \in \mathbb{C} \mid |\tau_3| \le 1\}$. Using

$$(\widetilde{Z}\times\widetilde{S}_3)\cup_{id}(\partial\widetilde{Z}\times\widetilde{D}_3) = (\widetilde{D}_1\times\widetilde{S}_2\times\widetilde{S}_3)\cup_{id}(\widetilde{S}_1\times\widetilde{S}_2\times\widetilde{D}_3) \simeq$$
$$[(\widetilde{D}_1\times\widetilde{S}_3) \cup_{id} (\widetilde{S}_1\times\widetilde{D}_3)] \times \widetilde{S}_2$$

we see that X is isomorphic to $S^3\times S^1$ and we can consider $M[C_1]$ as a 4-manifold obtained from $S^3\times S^1$ by surgery along a smooth circle, say ℓ_1, embedded in $S^3\times S^1$. Without loss of generality we can assume that $C_2 \subset (\widetilde{Y}-\partial\widetilde{Y})$. Thus the image of C_2 in X is well defined. Denote the corresponding smooth circle in $S^3\times S^1$ ($\simeq$ X) by ℓ_2. Now we can consider N as a 4-manifold obtained from $S^3\times S^1$ by surgeries along ℓ_1 and ℓ_2 ($\ell_1 \cap \ell_2 = \emptyset$). Let $x \in S^3$, $\overline{\ell} = x\times S^1$, ℓ_1 (resp. ℓ_2) be homologically equivalent to $n_1\overline{\ell}$ (resp. $n_2\overline{\ell}$). Since $\pi_1(N) = 0$ we have that either $n_1 = 0$, $n_2 \neq 0$ or $n_1 \neq 0$, $n_2 = 0$ or $n_1 \neq 0$, $n_2 \neq 0$, n_1, n_2 are relatively prime. Consider the third case, that is, $n_1 \neq 0$, $n_2 \neq 0$, n_1, n_2 are relatively prime. We can assume $n_1 > 0$, $0 < n_2 \leq n_1$. Let $n_1 = n_2 q + r$, where $r, q \in \mathbb{Z}$, $0 \leq r < n_2$, $\ell(r)$ be a smooth circle in $S^3\times S^1$ homologically equivalent to $r\overline{\ell}$ and such that $\ell(r) \cap \ell_1 = \emptyset$, $\ell(r) \cap \ell_2 = \emptyset$, $X(n_2)$ be a 4-manifold obtained from X by surgery along ℓ_2 and $\ell_1', \ell'(r)$ be the images of $\ell_1, \ell(r)$ in $X(n_2)$. It is clear that $\pi_1(X(n_2)) = \mathbb{Z}/n_2\mathbb{Z}$ and that $\ell_1', \ell'(r)$

are homologically equivalent in $X(n_2)$ (that is, $\ell_1', \ell'(r)$ correspond to homotopically equivalent embeddings of S^1 in $X(n_2)$). Now we use the following remark of Wall (see [8], p. 135): For 1-manifolds in 4-manifolds every homotopy may be replaced by an isotopy. We see that we obtain N by performing surgery in $X(n_2)$ along $\ell'(r)$, that is, we obtain N from $S^3 \times S^1$ by surgeries along ℓ_2 and $\ell(r)$. Hence we can replace the pair (n_1, n_2) by (n_2, r). Repeating this process after finite number of steps we come to the pair $(n', 0)$. Thus we could assume from the beginning that our pair (n_1, n_2) is $(n', 0)$. Because $\pi_1(N) = 0$ we have $n' = 1$. But then $X(n') = X(1) \approx S^4$ and N is diffeomorphic to an S^2-bundle over S^2.

Q.E.D.

Lemma 14. Let $f: M \to S$ be a Kodaira Fibration. Suppose that $\pi_1(M) = 0$ and $\chi(M) = 1$. Then $M \# P$ is diffeomorphic to $2P \# 9Q$.

Proof. Let ν be the number of multiple fibers of $f: M \to S$. By Lemma 12, $\nu \leq 2$ and $f: M \to S$ is regular. If $\nu \leq 1$ then our Lemma follows from Lemma 10. Consider the case $\nu = 2$.

From Lemma 9 and Theorem 9 it follows that the Lefshetz fibration corresponding to $f: M \to S$ coincides with $f_0: M_0 \to S_0$. Let $\{c_1, c_2\}$, $\{E_1, E_2\}$, $\{(m_1, k_1), (m_2, k_2)\}$, $\{(\varphi_1, \psi_1), (\varphi_2, \psi_2)\}$ be defined for $f: M \to S$ as in Definition 8. From Definition 8a it follows now that we can consider M_o as union of $M - \bigcup_{j=1}^{2} f^{-1}(c_j)$ and $F(D_\epsilon)_1, F(D_\epsilon)_2$ where we identify $x \in f^{-1}(E_j - c_j)$, $j = 1,2$, with $\Lambda_{m_j,k_j} \psi_j(x) \in F^\cdot(D_\epsilon)_j$ (see Definition 8a). We can identify also S_o with S so that for $j = 1,2$ $f_o^{-1}(E_j)$ will be equal to $F(D_\epsilon)_j$.

Thus we have an identification of $f_o^{-1}(E_j)$ with $E_j \times (\mathbb{C}/(i,1))$, where $(i,1)$ means the group of automorphisms of $\mathbb{C}$ consisting of transformations

$$\zeta \longrightarrow \zeta + n_1 i + n_2, \qquad n_1, n_2 \in \mathbb{Z}.$$

Let e_{1j}, e_{2j} be the basis of homologies of $H_1(f_1^{-1}(c_j),\mathbb{Z})$ corresponding to the vectors i and 1 on $\mathbb{C}$. We shall prove that there exists a smooth path $\overline{\gamma}$ on S_o connecting c_1 and c_2 and such that $\overline{\gamma}$ does not contain critical values of $f_o: M_o \longrightarrow S_o$ and the isomorphism $\Theta_{\overline{\gamma}}: H_1(f^{-1}(c_1),\mathbb{Z}) \longrightarrow H_1(f^{-1}(c_2),\mathbb{Z})$, corresponding to $\overline{\gamma}$, has the following property:

$$\Theta_{\overline{\gamma}}(e_{11}) = e_{12}, \quad \Theta_{\overline{\gamma}}(e_{21}) = e_{22}.$$

Let $\tilde{x} = \left\|\begin{matrix}1 & 0\\ -1 & 1\end{matrix}\right\|$, $\tilde{y} = \left\|\begin{matrix}1 & 1\\ 0 & 1\end{matrix}\right\|$. It is well known that the group $SL(2,\mathbb{Z})$ is generated by the matrices $\tilde{x}$ and $\tilde{y}$. For any $A \in SL(2,\mathbb{Z})$ denote by $d(A)$ the minimum of the lengths of all words in the alphabet $\tilde{x},\tilde{y},\tilde{x}^{-1},\tilde{y}^{-1}$ which correspond to A. Let Ω be the set of all smooth paths γ which connect c_1 with c_2 and do not contain critical values of $f_o: M_o \longrightarrow S_o$. For any $\gamma \in \Omega$ let

$\theta_\gamma: H_1(f^{-1}(c_1),\mathbb{Z}) \to H_1(f^{-1}(c_2),\mathbb{Z})$ be the isomorphism corresponding to γ and A_γ be the element of $SL(2,\mathbb{Z})$ corresponding to θ_γ and to the bases e_{11}, e_{21} of $H_1(f^{-1}(e_1),\mathbb{Z})$ and e_{12}, e_{22} of $H_1(f^{-1}(c_2),\mathbb{Z})$. Denote by $d_\gamma = d(A_\gamma)$. Let $\tilde{d} = \min_{\gamma\in\Omega} d_\gamma$ and $\tilde{\gamma}$ be an element of Ω with $d_{\tilde{\gamma}} = \tilde{d}$. We claim that $d_{\tilde{\gamma}} = 0$ (that is, we can take $\bar{\gamma} = \tilde{\gamma}$). Suppose that $d_{\tilde{\gamma}} > 0$. Take a closed 2-disk D in S_o such that $\tilde{\gamma} \subset \mathrm{int}(D)$ and D does not contain critical values of $f_o: M_o \to S_o$. From Lemma 8 it follows that we can find two critical values, say a_1, a_2, of $f_o: M_o \to S_o$ and two smooth paths γ_1, γ_2 on S_o such that $\gamma_1 \cap \gamma_2 = c_1$, γ_1 (corresp. γ_2) connects c_1 with a_1 (corresp. a_2), γ_1 (corresp. γ_2) does not contain critical values of $f_o: M_o \to S_o$ different from a_1 (corresp. a_2) and if θ_1 (corresp. θ_2) denotes the automorphism of $H_1(f^{-1}(e_1),\mathbb{Z})$ corresponding to γ_1 (corresp. γ_2) then the unimodular 2-matrix $A(\theta_1)$ (corresp. $A(\theta_2)$) which corresponds to θ_1 (corresp. θ_2) and to the basis e_{11}, e_{21} of $H_1(f^{-1}(c_1),\mathbb{Z})$ is equal to $\tilde{x}$ (corresp. $\tilde{y}$). It is easy to see that we can assume that each of γ_1, γ_2 intersects ∂D only in one point. Hence we can change γ_1, γ_2 so that we will have $\gamma_1 \cap \tilde{\gamma} = c_1$, $\gamma_2 \cap \tilde{\gamma} = c_1$. Let $W(A_{\tilde{\gamma}})$ be a word in the alphabet $\tilde{x}, \tilde{y}, \tilde{x}^{-1}, \tilde{y}^{-1}$ which corresponds to $A_{\tilde{\gamma}}$ and has the minimal length. We can assume that there exists a small circle s on S_o with the center c_1 and such that each of $\gamma_1, \gamma_2, \tilde{\gamma}$ intersects s

only in one point. Let $b_1 = \gamma_1 \cap s$, $b_2 = \gamma_2 \cap s$, $b = \tilde{\gamma} \cap s$. Let us say that we are in the case I (corresp. II) if the triple (b_1,b,b_2) corresponds to positive (corresp. negative) rotation of s Let α be the first letter in $W(A_{\tilde{\gamma}})$ from the left. (We shall write here the composition of transformations from the left to the right, as multiplication of matrices!). We say that we are in the case (I,α') (corresp. (II,α')) where α' is equal to one of $\tilde{x},\tilde{y},\tilde{x}^{-1},\tilde{y}^{-1}$, if we are in the case I (corresp. II) and $\alpha = \alpha'$. Now for each of our eight cases we construct new path $\tilde{\gamma}'$ as it is shown in Fig. 10. It is easy to verify directly that $A_{\tilde{\gamma}'} = \alpha^{-1}A_{\tilde{\gamma}}$. Hence $d_{\tilde{\gamma}'} = d(A_{\tilde{\gamma}'}) < d(A_{\tilde{\gamma}}) = d_{\tilde{\gamma}}$. We obtain a contradiction with the minimality of $d_{\tilde{\gamma}}$. Thus $d_{\tilde{\gamma}} = 0$ and we take $\bar{\gamma} = \tilde{\gamma}$.

Let $\underline{D}$ be a closed 2-disk on S_o such that $E_j \subset \mathrm{int}(\underline{D})$, $j = 1,2$, $\bar{\gamma} \subset \mathrm{int}(\underline{D})$. Using $\theta_{\bar{\gamma}}(e_{11}) = e_{12}$, $\theta_{\bar{\gamma}}(e_{21}) = e_{22}$ we can construct a trivialization of $f_o|_{f_o^{-1}(\underline{D})}: f_o^{-1}(\underline{D}) \to \underline{D}$, say

$$\begin{array}{ccc} f_o^{-1}(\underline{D}) & \xrightarrow{\ \underline{\Phi}\ } & \underline{D} \times T^2 \\ {\scriptstyle f_o}\downarrow & & \downarrow{\scriptstyle pr} \\ \underline{D} & \xrightarrow[\ id\ ]{} & \underline{D} \end{array} ,$$

where $T^2 = \mathbb{C}/(i,1)$, such that for $f_o^{-1}(E_j)$, $j = 1,2$, $\underline{\Phi}$ coincides with our previous identification $f_o^{-1}(E_j) = E_j \times (\mathbb{C}/(i,1))$.

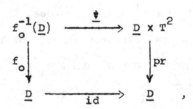

Fig. 10

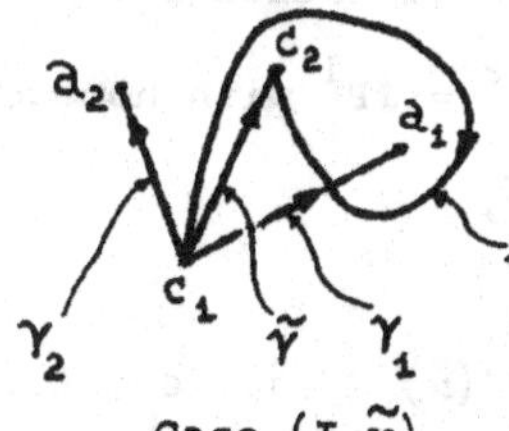

Case $(I,\tilde{x})$

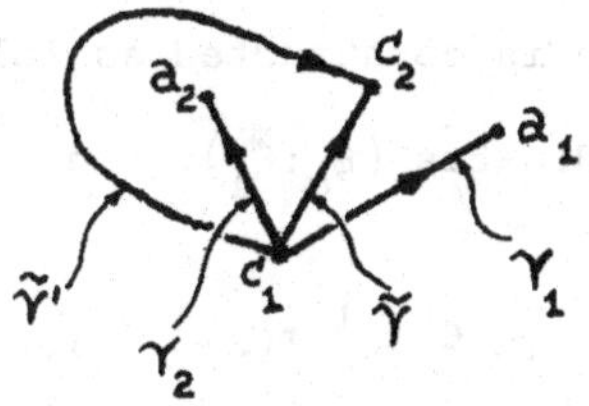

Case $(I,\tilde{y})$

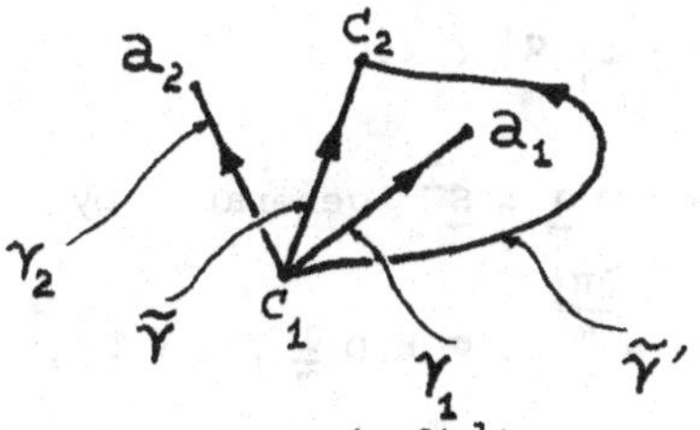

Case $(I,\tilde{x}^{-1})$

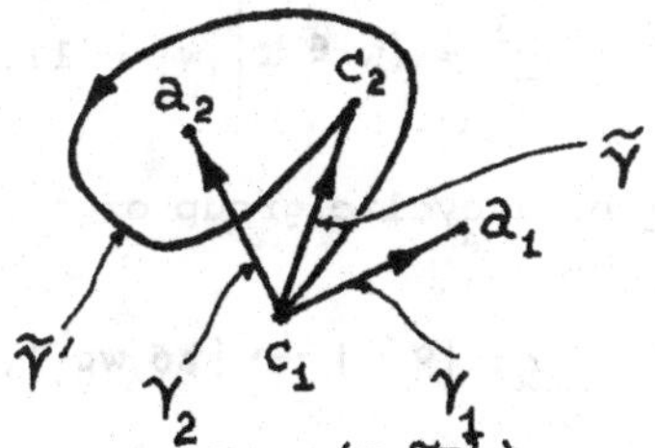

Case $(I,\tilde{y}^{-1})$

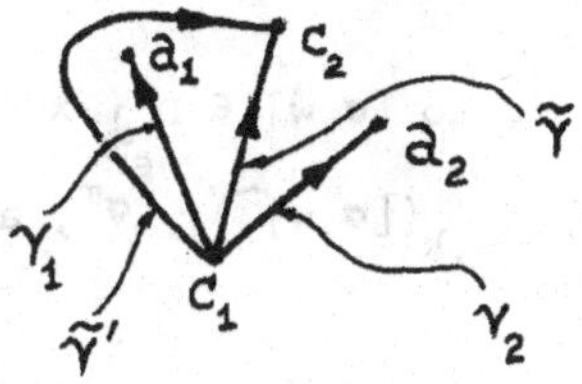

Case $(II,\tilde{x})$

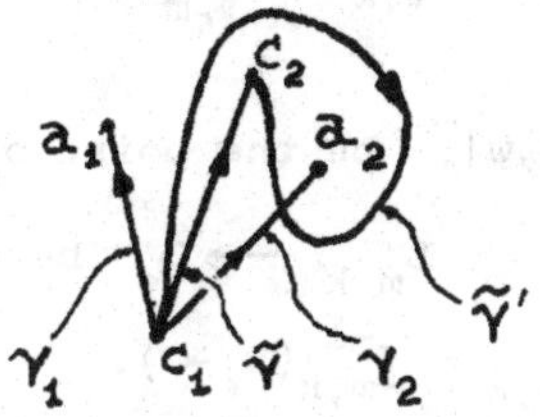

Case $(II,\tilde{y})$

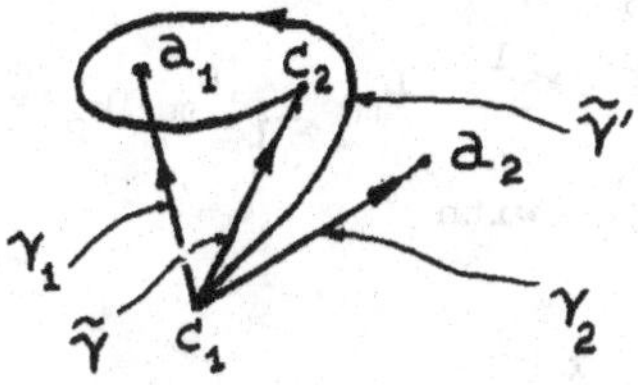

Case $(II,\tilde{x}^{-1})$

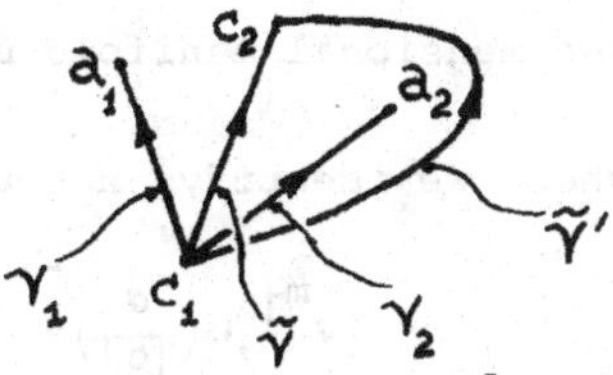

Case $(II,\tilde{y}^{-1})$

Now it is easy to verify that $f: M \longrightarrow S$ is isomorphic to the direct sum of $f_o: M_o \longrightarrow S_o$ and a Kodaira fibration $f_2: M_2 \longrightarrow S$ which is constructed as follows: We take $S^2 = \mathbb{C}P^1$ with homogeneous coordinates $(\xi_o:\xi_1)$. Let $\tau = \frac{\xi_1}{\xi_o}$, $\tau' = \frac{\xi_o}{\xi_1}$,

$$E = \{x \in \mathbb{C}P^1 \mid \tau(x) \leq \epsilon\}, \quad E' = \{x \in \mathbb{C}P^1, \tau'(x) < \epsilon\}, \quad \epsilon < 1.$$

We shall use the following notations: Let $k,m \in \mathbb{Z}$,

$$\underline{S}^1 = \{w \in \mathbb{C} \mid |w| = 1\}, \qquad D_\epsilon = \{\sigma \in \mathbb{C} \mid |\sigma| < \epsilon\},$$

$\underline{G}$ be a cyclic group of automorphisms of $D_{\epsilon^{\frac{1}{m}}} \times \underline{S}^1$ generated by

$$\underline{g}: [\sigma,w] \longrightarrow [\rho\sigma, w\rho^k], \text{ where } \rho = e^{\frac{2\pi i}{m}}, \sigma \in D_{\epsilon^{\frac{1}{m}}}, w \in \underline{S}^1,$$

$$L_{m,k} = (D_{\epsilon^{\frac{1}{m}}} \times \underline{S}^1)/\underline{G},$$

$[\sigma,w]^\sim$ be the point of $L_{m,k}$, corresponding to $[\sigma,w] \in D_{\epsilon^{\frac{1}{m}}} \times \underline{S}^1$, $\underline{f}_{m,k}: L_{m,k} \longrightarrow D_\epsilon$ be the map defined by $\underline{f}_{m,k}([\sigma,w]^\sim) = \sigma^m$, and $L^{\cdot}_{m,k} = \underline{f}^{-1}_{m,k}(D_\epsilon - 0)$.

Let $c_o, c_\infty \in \mathbb{C}P^1$ be defined by $\tau(c_o) = 0$, $\tau'(c_\infty) = 0$, and $E^{\cdot} = E - c_o$, $E'^{\cdot} = E' - c_\infty$, $U = \mathbb{C}P^1 - (c_o \cup c_\infty)$. Define a 3-dimensional manifold L as the union of $U \times \underline{S}^1$, L_{m_1,k_1}, L_{m_2,k_2}, where we identify $x \in L^{\cdot}_{m_1,k_1}$, $x = [\sigma,w]^\sim$, with

$$[\sigma^{m_1}, w(\tfrac{\sigma}{|\sigma|})^{-k_1}] \in E^{\cdot} \times \underline{S}^1 \subset U \times \underline{S}^1$$

and $x' \in L_{m_2,k_2}$, $x' = [\sigma', w']^{\sim}$ with

$[(\sigma')^{m_2}, w'(\frac{\sigma'}{|\sigma'|})^{-k_2}] \in E! \times \underline{S}^1 \subset U \times \underline{S}^1$. Let $\underline{f}: L \to \mathbb{C}P^1$

be defined by

$$\underline{f}(x) = \begin{cases} pr_U(x) & \text{when } x \in U \times \underline{S}^1, \\ \underline{f}_{m_1,k_1}(x) & \text{when } x \in L_{m_1,k_1}, \\ \underline{f}_{m_2,k_2}(x) & \text{when } x \in L_{m_2,k_2} \end{cases}$$

Now we take $M_2 = L \times S^1$ and $f_2 = \underline{f} \cdot (pr_L)$.

By Lemma $11_{ii)}$ we have that $M \# P$ is diffeomorphic to $P \# 8Q \# M_2^{\approx}$ where $M_2^{\approx}$ is a 4-manifold obtained from M_2 by surgeries along certain disjoint smooth circles s_b and s_c constructed in Lemma 11. This construction (see the formulation of Lemma 11) is such that we can take $s_b = p \times S^1$, $s_c = \ell \times q$, where $p \in L$, $q \in S^1$ and ℓ is a smooth circle in L. Since $\pi_1(M) = 0$ we have $\pi_1(M \# P) = 0$ and then $\pi_1(M_2^{\approx}) = 0$. Now it is easy to see that L is diffeomorphic to a lense space and because $\pi_1(M_2^{\approx}) = 0$ the circle ℓ must be a representative for a generator of $\pi_1(L)$. Using Lemma 13, we obtain that $M_2^{\approx}$ is diffeomorphic to an S^2-bundle over S^2. Because the intersection form of $P \# 8Q$ is odd, we have that $P \# 8Q \# M_2^{\approx}$ is diffeomorphic to $P \# 8Q \# P \# Q$ (see [8]). Thus $M \# P \approx 2P \# 9Q$.

Q.E.D.

Theorem 11. Let $f: M \longrightarrow S$ be a Kodaira fibration with ν multiple fibers such that $\pi_1(M) = 0$. Then $\chi(M) = \frac{e(M)}{12}$ is a positive integer and the following is true:

i) if $\chi(M) = 1$ and $\nu \neq 2$ then M is diffeomorphic to $P \# gQ$;

ii) in all cases $M \# P$ is diffeomorphic to

$$2\chi(M)P \# (10\chi(M)-1)Q.$$

Proof. i) follows from Lemma 12 and Lemma 10. Because $\pi_1(M) = 0$ we have that $\chi(M)$ is a positive integer by Lemma 12 and Lemma 9.

ii) Using Lemma 14 we see that it would be enough to prove the following inductive

Statement (*). Suppose that $\chi(M) > 1$ and that for any Kodaira fibration $f': M' \longrightarrow S'$ with $\pi_1(M') = 0$ and $\chi(M') < \chi(M)$ we have that $M' \# P$ is diffeomorphic to $2\chi(M')P \# (10\chi(M')-1)Q$. Then $M \# P$ is diffeomorphic to $2\chi(M)P \# (10\chi(M)-1)Q$.

Proof of Statement (*). From Lemma 12 we know that $\nu \leq 2$ and that the Lefshetz fibration of 2-toruses $\tilde{f}: \tilde{M} \longrightarrow \tilde{S}$ corresponding to $f: M \longrightarrow S$ is regular. We have also $e(M) = e(\tilde{M})$ (Lemma 9). Since $\chi(M) > 1$, we obtain from Theorem 9 that there exists a regular Lefshetz fibration of 2-toruses $\tilde{f}_2: \tilde{M}_2 \longrightarrow \tilde{S}_2$ such that $\tilde{f}: \tilde{M} \longrightarrow \tilde{S}$ is isomorphic to the direct sum of $f_o: M_o \longrightarrow S_o$ and $\tilde{f}_2: \tilde{M}_2 \longrightarrow \tilde{S}_2$. Using Definitions 8, 8a and 10 we can construct a

Kodaira fibration $f_2: M_2 \longrightarrow S_2$ with ν multiple fibers such that $\tilde{f}_2: \tilde{M}_2 \longrightarrow \tilde{S}_2$ is the Lefshetz fibration corresponding to $f_2: M_2 \longrightarrow S_2$ and $f: M \longrightarrow S$ is isomorphic to the direct sum of $f_o: M_o \longrightarrow S_o$ and $f_2: M_2 \longrightarrow S_2$. By Lemma 11 ii) we have that $M \# P$ is diffeomorphic to $P \# 8Q \# \approx{M}_2$ where $\approx{M}_2$ is obtained from M_2 by surgeries along two smooth disjoint circles embedded in M_2. Let $S^2 \underset{\sim}{\times} S^2$ be either $S^2 \times S^2$ or $P \# Q$. Applying Lemma 12 to $f: M \longrightarrow S$ and $f_2: M_2 \longrightarrow S_2$ we obtain $\pi_1(M_2) = 0$ (we use $\pi_1(M) = 0$). Then by the results of Wall (see [8]) we get that $\approx{M}_2$ is diffeomorphic to $M_2 \# 2(S^2 \underset{\sim}{\times} S^2)$. Hence $M \# P$ is diffeomorphic to $P \# 8Q \# M_2 \# 2(S^2 \underset{\sim}{\times} S^2)$. Since the intersection form of $P \# 8Q \# M_2$ is of odd type we obtain (again referring to [8]) that $(P \# 8Q \# M_2) \# 2(S^2 \underset{\sim}{\times} S^2) \approx (P \# 8Q \# M_2) \# 2(P \# Q) \approx M_2 \# (3P \# 10Q)$. It is clear that $\chi(M) = \chi(M_2) + \chi(M_o)$, that is, $\chi(M_2) = \chi(M) - 1$. Using the supposition of the induction we have that

$$M_2 \# P \approx 2\chi(M_2)P \# 10\chi(M_2)-1)Q.$$

Hence

$$M \# P \approx M_2 \# (3P \# 10Q) \approx (M_2 \# P) \# (2P \# 10Q) =$$

$$2\chi(M_2)P \# (10\chi(M_2)-1)Q \# 2P \# 10Q \approx$$

$$2\chi(M)P \# (10\chi(M)-1)Q. \qquad \text{Q.E.D.}$$

§4. Topology of simply-connected elliptic surfaces.

Theorem 12. Let V be a simply-connected elliptic surface. Then V is almost completely decomposable.

Proof. Without loss of generality we can assume that V is a minimal elliptic surface, that is, there exists a holomorphic map $f: V \longrightarrow \Delta$, where Δ is a compact Riemann surface such that for generic $x \in \Delta$, $f^{-1}(\Delta)$ is an elliptic curve and there are no exceptional curves in the fibers of f. From Theorems 8 and 8a it follows that we can assume that all singular fibers of $f: V \longrightarrow \Delta$ are of type ${}_mI_o$ or I_1.

It is easy to see then that $f: V \longrightarrow \Delta$ will be a Kodaira fibration. Now the theorem follows from Theorem 11. Q.E.D.

Corollary. Let V_1, V_2 be simply-connected elliptic surfaces with $b_2(V_1) = b_2(V_2)$, $\tau(V_1) = \tau(V_2)$ ($\tau(V)$ is the signature of V). Then $V_1 \# P$ is diffeomorphic to $V_2 \# P$.

APPENDIX II

R. Livne. A theorem about the modular group.

Let G be a group, and $(a_1,\cdots,a_n)$ be an n-tuple in G. Let T be the set of all n-tuples $(b_1,\cdots,b_n)$ in G which satisfy:

1) $b_1b_2\ldots b_n = a_1a_2\ldots a_n$;
2) there is a permutation σ of $1,\cdots,n$ such that b_i is conjugate to $a_{\sigma(i)}$.

For $1 \leq i \leq n-1$ define transformations R_j on T:

$$R_j(b_1,\cdots,b_{j-1},b_j,b_{j+1},b_{j+2},\cdots) = (b_1\ldots b_{j-1},b_{j+1},b_{j+1}^{-1}b_jb_{j+1},b_{j+2},\cdots).$$

It is clear that R_j maps T into itself: the conditions 1), 2) are still satisfied after their application. Moreover, R_j are bijective. In fact, the inverses are given by

$$R_j^{-1}(b_1,\ldots,b_{j-1},b_j,b_{j+1},b_{j+2},\ldots) = (b_1,\ldots,b_{j-1},b_jb_{j+1}b_j^{-1},b_j,b_{j+2},\cdots).$$

Both R_j and R_j^{-1} will be called "the elementary transformations".

The analysis of "global monodromy" in complex geometry motivates the following algebraic problem:

Let C be the group of transformations of T which are equal to finite sequences of elementary transformations. Describe the action of C on T. (For example, when it is transitive?)

The case when G is a free group and $\{a_1,\cdots,a_n\}$ are free generators of G was considered by E. Artin in [21]. He proved that in this case C acts transitively on T.

Let $a_1\cdot\ldots\cdot a_n = A$. Applying $R_{n-1}\cdot\ldots\cdot R_2\cdot R_1$ to $(b_1,\cdots,b_n) \in T$ we get $(b_2,\cdots,b_n,A^{-1}b_1A)$. Now if A is in the center of the group G (especially if it is 1) we see that C contains any cyclic shift on $(b_1,\cdots,b_n)$.

Now let G be the modular group $\Gamma \simeq \mathbb{Z}_3*\mathbb{Z}_2$, which we present as $\{a,b \mid a^3 = b^2 = 1\}$. Any element in G can be expressed as a word in a and b. Each element $g \in G$ has a unique presentation as $t_1\cdot\ldots\cdot t_k$, where each t_i is a,a^2 or b, and successive t_i's cannot be two b's or two powers of a. We call such a presentation $g = t_1 \ldots t_k$ "reduced" and define $\ell(g) = k$, the length of g.

Let $s_o = a^2b$, $s_1 = aba$, $s_2 = ba^2$. They clearly satisfy:

a) $s_2s_1s_os_2s_1s_o = 1$.

b) They are conjugate to each other (by powers of a).

It is clear that for an element g in reduced form $t_1\cdots t_k$, the reduced form of the inverse g^{-1} is $t_k' \cdots t_1'$ where

$$t_i' = \begin{cases} b & \text{if } t_i = b, \\ a^2 & \text{if } t_i = a, \\ a & \text{if } t_i = a^2. \end{cases}$$

If we conjugate s_o or s_2 by en element t of length 1, we end up with one of the s_i. Therefore, any conjugate of the s_i is either one of them, and then we call it "short" or else it is $Q^{-1}abaQ = Q^{-1}s_1Q$, where Q is expressed in reduced form $t_1\cdots t_k$, Q^{-1} is written as $t_k'\cdots t_1'$ and $t_1 = b$.

In this case we say this conjugate is "long".

<u>Theorem</u>. Let $g_1,\cdots,g_n$ be conjugates of the s_1, such that

$$g_1\cdot\ldots\cdot g_n = 1.$$

Then, by successive application of elementary transformations the n-typle $(g_1,\cdots,g_n)$ can be transformed to an n-tuple $(h_1,\cdots,h_n)$ with each h_i short.

<u>Proof</u>. Express each of the long g_i as $Q_i^{-1}s_1Q_i$, and define

$$L(g_1,\cdots,g_n) = \sum \ell(Q_i).$$

We carry the proof by induction on $L(g_1,\cdots,g_n)$. If $L(g_1,\cdots,g_n) = 0$, all the g's are short, as required. For the induction, we need the following

<u>Assertion</u>. For some i, $1 \leq i \leq n-1$, we have $\ell(g_i\cdot g_{i+1}) < \max(k,\ell)$ where $k = \ell(g_i)$, $\ell = \ell(g_{i+1})$.

<u>Proof</u>. Assume the contrary, that for all i, $1 \leq i \leq n-1$, we have $\ell(g_ig_{i+1}) \geq \ell(g_i),\ell(g_{i+1})$. We shall show that the

expression $g_1 \cdots g_n$ cannot reduce to 1, a contradiction.

Let i be fixed and $k = \ell(g_i)$, $\ell = \ell(g_{i+1})$. Write $g_i = t_k \cdots t_1$, $g_{i+1} = \tilde{t}_1 \cdots \tilde{t}_\ell$, where $\ell(t_j) = \ell(\tilde{t}_{j'}) = 1$, $j = 1,\ldots,k$, $j' = 1,\ldots,\ell$, (that is, in reduced form).

The reduced form of $g_i g_{i+1}$ is then either $t_k \ldots t_{m+1} r \tilde{t}_{m+1} \ldots \tilde{t}_\ell$ or $t_k \ldots t_{m+1} r$ or $r \tilde{t}_{m+1} \ldots \tilde{t}_\ell$, where $\ell(r) \leq 1$.

There are two cases:

1) $\ell(r) = 0$, that is, $r = 1$. If $g_i g_{i+1} = t_k \ldots t_{m+1}$ or $g_i g_{i+1} = \tilde{t}_{m+1} \ldots \tilde{t}_\ell$, we have $\ell(g_i g_{i+1}) = |\ell - k| < \max(k,\ell)$. Contradiction. Hence

$$g_i g_{i+1} = t_k \ldots t_{m+1} \tilde{t}_{m+1} \ldots \tilde{t}_\ell.$$

Suppose $m > 0$. Since $t_k \ldots t_{m+1} \tilde{t}_{m+1} \ldots \tilde{t}_\ell$ is the reduced form of $g_i g_{i+1}$, one of the $t_{m+1}, \tilde{t}_{m+1}$ must be equal to a or a^2 and the second must be equal to b. But then one of $t_m, \tilde{t}_m$ must be equal to b and because $t_m \tilde{t}_m = 1$ we have $t_m = \tilde{t}_m = b$. That means that either $t_{m+1} = t_m = b$ or $\tilde{t}_{m+1} = \tilde{t}_m = b$. Contradiction. Thus $m = 0$.

2) $\ell(r) = 1$, that is, $r = a$ or a^2, $t_{m+1} = \tilde{t}_{m+1} = b$. In this case $t_j = \tilde{t}_j^{-1}$ for $1 \leq j \leq m-1$ and $t_m = \tilde{t}_m = a^2$ or a. We have $k,\ell \leq \ell(g_i g_{i+1}) = k+\ell-2m+1$, hence $k,\ell \geq 2m-1$. If k is even, g_i must be s_o or s_2, since all other conjugates of the s_1 are of

odd length. Therefore $k = 2$, $m = 1$, $t_1 = a^2$ or a. Hence $g_i = s_2$. We have $k \geq 2m$ in this case.

Similarly, if ℓ is even, then $\ell \geq 2m$ and $g_{i+1} = s_o$. If k is odd, write $g_i = Q_i^{-1}abaQ_i$, where $q = \ell(Q_i) \geq 0$. We have $k = 2q+3$.

Suppose $2m-1 = k$. Then $m = q+2$ and we have $t_m = b$ which contradicts $t_m = a^2$ or a. Therefore $k > 2m-1$ and because k is odd, we have $k \geq 2m+1 > 2m$. Similarly, if ℓ is odd, $\ell > 2m$.

Now denote for each $i = 1,2,\cdots,n$, $k_i = \ell(g_i)$ and write each g_i, $i = 1,2,\cdots,n$, in the reduced form as follows:

$$g_i = t_{k_i}^{(i)} \cdots t_1^{(i)} \equiv \tilde{t}_1^{(i)} \cdots \tilde{t}_{k_i}^{(i)},$$

that is, for any $j = 1,2,\cdots,k_i$, $t_j^{(i)} = \tilde{t}_{k_i-j+1}^{(i)}$. Denote r,m considered above (for the pair g_i,g_{i+1}) respectively $r_{i,i+1}$, $m_{i,i+1}$. Let $r_{o,1} = r_{n,n+1} = 1$, $m_{oi} = m_{n,n+1} = 0$.
Our above consideration shows the following:

a) If $m_{i,i+1} \neq 0$ then $r_{i,i+1} = a$ or a^2 and $t_{m_{i,i+1}+1}^{(i)} = \tilde{t}_{m_{i,i+1}+1}^{(i+1)} = b$.

If $m_{i,i+1} = 0$, then $r_{i,i+1} = 1$.

b) If k_i is even and $m_{i,i+1} \neq 0$, then $g_i = ba^2 = s_2$ and $m_{i-1,i} = 0$, $r_{i-1,i} = 1$. If $m_{i-1,i} \neq 0$ (k_i even), then $g_i = a^2b = s_o$ and $m_{i,i+1} = 0$, $r_{i,i+1} = 1$. We see here that $m_{i-1,i}+m_{i,i+1} \leq k_i-1$.

c) If k_i is odd, then $m_{i-1,i} \leq \frac{k_i-1}{2}$, $m_{i,i+1} \leq \frac{k_i-1}{2}$, that is,

$m_{i-1,i} + m_{i,i+1} \leq k_i-1$. Now from $m_{i-1,i}+m_{i,i+1} \leq k_i-1$ it follows that $m_{i-1,i}+1 \leq k_i-(m_{i,i+1}+1)+1$, that is, in the reduced form of g

$\tilde{t}^{(i)}_{m_{i-1,i}+1}$ is either on the left side from

$$t^{(i)}_{m_{i,i+1}+1} = \tilde{t}_{k_i-(m_{i,i+1}+1)+1}$$

or $\tilde{t}^{(i)}_{m_{i-1,i}+1}$ coincides with $t^{(i)}_{m_{i,i+1}+1}$. Now we can write

$$(*) \qquad g_1 \cdots g_n = \prod_{i=1}^{n} (\tilde{t}^{(i)}_{m_{i-1,i}+1} \cdots \tilde{t}^{(i)}_{k_i-(m_{i,i+1}+1)+1})r_{i,i+1}.$$

Note that if $r_{i,i+1} = 1$ $(i \leq n-1)$, then one of $\tilde{t}^{(i)}_{k_i-(m_{i,i+1}+1)+1}$, $\tilde{t}^{(i+1)}_{m_{i,i+1}+1}$ is equal to a^2 or a and the second is equal to b. If $r_{i,i+1} \neq 1$, then $r_{i,i+1} = a^2$ or a and $\tilde{t}^{(i)}_{k_i-(m_{i,i+1}+1)+1} = \tilde{t}^{(i+1)}_{m_{i,i+1}+1} = b$. We see that the right side in (*) is the reduced form of $g_1 \ldots g_n$. Hence $g_1 \ldots g_n \neq 1$. Contradiction. The assertion is proved.

We return to the proof of the theorem:

If $\ell(g_i g_{i+1}) < \ell(g_i)$ we shall show that application of R_i or R_i^{-1} will reduce L if not both g_i and g_{i+1} are short, and similarly if $\ell(g_i g_{i+1}) < \ell(g_{i+1})$.

Assuming, as we might, $\ell(g_i) \geq \ell(g_{i+1})$ we show that $\ell(g_i) > \ell(g_{i+1})$ (if $\ell(g_i) \leq \ell(g_{i+1})$ it will follow that $\ell(g_i) < \ell(g_{i+1})$). In fact this is clear if g_{i+1} is short. If it is long, then $g_i = Q_i^{-1} aba\, Q_i$ and $g_{i+1} = Q_{i+1}^{-1} aba\, Q_{i+1}$. If $\ell(Q_i) = \ell(Q_{i+1})$, the b's do not cancel, and $\ell(g_i g_{i+1}) > \ell(g_i), \ell(g_{i+1})$. Thus $\ell(g_i) > \ell(g_{i+1})$. Now we have $\ell(g_{i+1}^{-1} g_i g_{i+1}) < \ell(g_i)$. Applying R_i we evidently will reduce $L = L(g_1, \cdots, g_n)$. If $\ell(g_i) < \ell(g_{i+1})$ application of R_i^{-1} will reduce L. To finish the proof we must consider the case where both g_i and g_{i+1} are short. In this case, (g_i, g_{i+1}) is (s_1, s_o), (s_o, s_2) or (s_2, s_1) and all these possibilities can be transformed to each other by R_i and R_i^{-1}. Now if one of g_j, $j = 1,2,\cdots,n$, is long, say $g_{j'}$ for $j' > i+1$, take the smallest such j'.

Consider the sequence $y_1 = g_i$, $y_2 = g_{i+1}, \cdots, y_u = g_{j'-1}$ where $u = j' - i$. Let us prove by induction that for any u', $2 \leq u' \leq u$, $(y_1, \cdots, y_{u'})$ can be transformed by a finite sequence of elementary transformations to $(y_1', \cdots, y_{u'}')$, where $y_{u'}'$ is any of s_o, s_1, s_2. For $u' = 2$ it is clear. Suppose it is true for some $u' < u$.

Define a function $v\colon (0,1,2) \longrightarrow (0,1,2)$ by $v(0) = 1$, $v(2) = 0$, $v(1) = 2$. Now let $y_{u'+1} = s_\tau$, $\tau \in (0,1,2)$. We can transform $(y_1, \cdots, y_{u'}, y_{u'+1})$ to $(y_1', \cdots, y_{u'}', y_{u'+1})$ with

$y'_{u'} = s_{v(\tau)}$. Now the pair $y'_{u'}, y_{u'+1}$ is (s_1, s_o), (s_o, s_2) or (s_2, s_1) and all these possibilities can be transformed to each other by elementary transformations. The inductive statement is proved.

We see that we can assume that $g_{j'-1}$ is any of s_o, s_1, s_2. But for one such s_τ, $\tau = 0,1,2$, $\ell(s_\tau g_{j'}) < \ell(g_{j'})$. We now apply R_j.

The same arguments work if there exists a long $g_{j'}$ with $j' < i$. The theorem is proved.

Bibliography

[1] Wall, C.T.C., "On simply-connected 4-manifolds", J. London Math. Soc. 39 (1964), 141-149.

[2] Mandelbaum, R., Moishezon, B., "On the topological structure of non-singular algebraic surfaces in $\mathbb{C}P^3$", Topology, N1 (1976), 23-40.

[3] Bombieri, E., "Canonical models of surfaces of general type", Publ. Math. IHES, No. 42 (1973), 171-219.

[4] Mandelbaum, R., Moishezon, B., "On the topological structure of simply-connected algebraic surfaces", Bulletin of AMS, Vol. 82, No. 5 (1976), 731-733.

[5] Mandelbaum, R., "Irrational connected sums and the topology of algebraic surfaces", to appear.

[6] Tyurina, G., "The resolution of singularities of flat deformations of double rational points", Funktsional Analiz i Prilozhen. (Functional analysis and its applications) 4: 1(1970), 77-83.

[7] Brieskorn, E., "Die Auflösung der rationalen Singularitäten holomopher Abbildungen", Math. Ann., 178 (1968), 255-270.

[8] Wall, C.T.C., "Diffeomorphisms of 4-manifolds", J. London Math. Soc. 39 (1964), 131-140.

[9] Artin, M., "On isolated rational singularities of surfaces", Amer. J. Math., 88 (1966), 129-136.

[10] Mandelbaum, R., Moishezon, B., "On the topology of simply-connected algebraic surfaces", to appear.

[11] Kodaira, K., "Pluricanonical systems on algebraic surfaces of general type", J. Math. Soc. Japan 20 (1968), 170-192.

[12] Severi, F., "Intorno ai punti doppi impropri di una superficie generale dello spazio a quattro dimensioni, e a' suoi punti tripli apparenti", Rendiconti del circolo mat. di Palermo, T. XV (1901), 33-51.

[13] Kodaira, K., "On compact analytic surfaces, II", Ann. of Math., 77 (1963), 563-626.

[14] Kodaira, K., "On the structure of compact complex analytic surfaces, I", Amer. J. Math., 86 (1964), 751-798.

[15] Shafarevich, I.R., "Algebraic surfaces", Proc. of Steklov Institute of Mathematics, Amer. Math. Soc., 1967.

[16] Wallace, A., "Homology theory on algebraic varieties", NY, Pergamon Press, 1958.

[17] Earle, C.J., Eells, J., "The diffeomorphism group of a compact Riemann Surface", Bull. AMS, Vol. 73, No. 4 (1967), 557-560.

[18] Kas, A., "On the deformation types of regular elliptic surfaces (preprint).

[19] Kodaira, K., "On homotopy K3 surfaces", Essays on Topology and Related Topics, Mėmoires dėdiės a Georges de Rham, Springer, 1970, pp. 58-69.

[20] Seifert, H., Threfall, W., "Lehrbuch der Topologie", Leipzig und Berlin, B.G. Teubner, 1934.

[21] Artin, E., "Theory of braids", Ann. Math. 48 (1947), 101-126.

Subject Index